VILLAGE REPUBLIC 2.0

Harnessing Bioresources through NextGen Biology and Artificial Intelligence

SANJAY KUMAR
with ARUN TIWARI

Sakāl Publications

Sakal Publications

VILLAGE REPUBLIC 2.0

Sakal Media Pvt. Ltd.
595, Budhwar Peth,
Pune - 411002, India

www.sakalpublications.com
sakalprakashan@esakal.com

First Edition: October 2025

ISBN No.: 978-93-49487-16-1

Proofreader: Anupama MA
Typesetting: Apurva Selukar

Printed in India by Sakal Media Pvt. Ltd.

To my parents,
Shri Mahendra Krishnan and Smt. Shakuntala Devi,
who brought me into this world and supported all my endeavours.
To **Prof. Suresh Kumar Sinha**,
who set my science on autopilot.
And to the **Faceless Villager**,
poor yet valiant guardian of soil and soul.

This book stands at the confluence of two great visions: **Mahatma Gandhi's** *Gram Swaraj*—the Village Republic rooted in self-reliance and dignity—and **Dr Kalam's** *PURA*—the modernisation of villages through technology, innovation, and inclusivity. By uniting these two streams of thought, the chapters ahead seek to chart a path where India's villages are not abandoned in pursuit of urban growth, but elevated into the very centres of prosperity, sustainability, and a model of integrated human development.

CONTENTS

Foreword 7

Introduction 9

Part I : The Road Travelled

1. Pollen on the Wind 15
2. The Lure of the New 29
3. Decoding Life 45
4. Into the Nano 71
5. Harvesting Possibilities 91

Part II : The Path Ahead

6. Blooming Prosperity 109
7. Wonder Grass 125
8. Value-addition 149
9. Bioplastics 166
10. Bioenergy 181

Epilogue 196

Acknowledgements 197

Advance praise for the book 200

FOREWORD

At NITI Aayog, we have consistently emphasised technology as a driver of socio-economic transformation. But transformation is meaningful only when technology is inclusive, accessible, and centred on people. Innovation cannot remain confined to laboratories or be available only to a privileged few. Its true value lies in reaching small and marginal farmers, empowering rural women, and fuelling the aspirations of village youth. Technology that deepens inequality cannot be called progress. But when it uplifts the weakest, enhances productivity at the grassroots, and restores dignity and opportunity to millions, it fulfils its highest purpose.

The vision outlined in this book aligns seamlessly with the ethos of inclusive growth, *Sabka Vikas*, which is integral to the goal of *Viksit Bharat 2047*. By proposing practical ways to harness bioresources through NextGen Biology and Artificial Intelligence (AI), it moves the debate beyond theory to action, empowering individuals. It illustrates how scientific advances can translate into tangible benefits: higher farm incomes, diversified livelihoods, reduced migration, and renewed confidence in the village as a place of opportunity. It reminds us that technology is not an end in itself but a means to build a more just, resilient, and sustainable society.

Equally significant is the emphasis this book places on youth and smallholders. True rural prosperity lies in enabling young people to innovate and thrive in their native villages, rather than being compelled to migrate to overcrowded cities. By placing farmers at the centre of value creation and nurturing rural entrepreneurship, we can restore dignity, hope, and pride to village life.

I commend the authors for presenting a compelling alternative to the dominant model of urban-led growth. Their ideas are not utopian, but both visionary and practical. Dr Sanjay Kumar and Prof. Arun Tiwari, distinguished scientists in their respective fields, bring to this work a rare combination of intellectual rigour and deep social commitment. Their insights make this book a timely and valuable contribution at a moment when rural India stands at a decisive crossroads.

I am confident that this book will inspire policymakers, researchers, entrepreneurs, and young people alike to work towards an India where the Village Republic is not an outdated concept of the past, but the foundation of a vibrant, inclusive, and sustainable future.

Dr Ramesh Chand
Eminent Economist and Member, Agriculture, NITI Aayog

September 2, 2025
New Delhi

INTRODUCTION

Mahatma Gandhi's idea of the Village Republic envisioned self-reliant communities where human dignity, nature, and livelihoods coexisted in harmony. Over the decades, however, this dream has been eroded. Industrial agriculture, exploitative markets, and relentless urbanisation have drained villages of both wealth and people. Smallholders remain trapped in cycles of debt and vulnerability, while rural youth migrate in search of opportunities that their own communities no longer provide. The result is a paradox: the land that sustains the nation often leaves its caretakers impoverished.

Today, a new possibility emerges. The combined power of NextGen Biology and Artificial Intelligence (AI) offers a pathway to resurrect Gandhi's vision—not as nostalgia, but as a future-ready blueprint. Rural India is abundant in bioresources—crops, livestock byproducts, medicinal and aromatic plants, microbial diversity, and agro-waste. When these resources are reimagined through the lens of biotechnology, they become raw materials for biofuels, bioplastics, nutraceuticals, organic fertilisers, and high-value biomaterials. What was once considered waste becomes a source of wealth.

Yet scientific innovation alone is not enough. AI serves as the great enabler, translating complex biological processes into accessible, farmer-friendly applications. Through predictive analytics, AI can guide smallholders toward the most viable value chains. Through intelligent networks, it can connect them directly with cooperatives, processors, and markets. Automating routine tasks can reduce drudgery while enhancing productivity. Most importantly, AI can democratise knowledge, ensuring that even the smallest farmer in the remotest village participates in the bio-based rural economy.

This book unfolds in two parts.

Part One is a biography of Dr Sanjay Kumar, tracing his remarkable journey into the world of NextGen Biology. From humble beginnings, he rose to become a pioneering scientist whose research carried him to high-altitude plains and the cold deserts of the Himalayas. There, he discovered how plants survive in extreme climates, unlocking molecular pathways that are not only vital for agriculture but also for human survival. These discoveries led to the development of phytomedicines, bridging plant science and human health. His story illustrates how perseverance, curiosity, and science can converge to yield insights with global significance.

Part Two turns from biography to vision. It argues for a complete transformation of the rural economy—expanding agriculture beyond the narrow confines of crop growing into an engine of livelihoods, value addition, and wealth creation. Here, the ideas of Dr A.P.J. Abdul Kalam resonate powerfully. He envisioned Providing Urban Amenities in Rural Areas (PURA), where villages are equipped with technology, infrastructure, and enterprise to match the opportunities of cities. Just as Gandhi spoke of Gram Swaraj and the Village Republic, Kalam articulated a complementary vision of rural empowerment through science and innovation. Together, these visions form the twin pillars of this book.

The co-author, Arun Tiwari, brings a rare perspective to this endeavour. As Dr Kalam's pupil, collaborator, and biographer, he not only documented the great scientist-president's dreams but also helped translate them into practice. In this book, Tiwari extends that mission—drawing on Kalam's ideals of innovation with compassion, technology with inclusivity, and science with service to humanity.

In this context, the book warns that a purely consumer-driven economy—one based on endless extraction, consumption, and profit—is bound to fail humanity. Instead, people must learn to live sustainable lives, and villages must become the wealth creators of the future by harnessing bioresources through NextGen Biology and Artificial Intelligence. Crucially, these tools must be developed in the public sector—to empower farmers and communities—not left to corporate profiteering.

This convergence of bioscience and AI carries the potential to reverse rural distress. Instead of leaving for crowded cities, young people can find purpose and prosperity in their villages—as innovators, technicians, and entrepreneurs in local bio-enterprises. Smallholders, once mere price-takers in commodity markets, can become value-creators and stakeholders in sustainable industries. Villages can transform from marginalised peripheries into hubs of innovation, inclusion, and resilience.

This book, Village Republic 2.0, explores how such a transformation can be achieved. It presents a vision where science, technology, and ethics converge to build a rural economy that is regenerative, prosperous, and future-ready. It asks a simple but profound question: can India reimagine its villages, not as relics of the past, but as laboratories of the future? The answer, as the chapters ahead will argue, is yes. By harnessing bioresources through NextGen Biology and Artificial Intelligence, we can make the village not just the heart of India's heritage, but also the engine of its future.

PART I
THE ROAD TRAVELLED

"The best way to find yourself is to lose yourself in the service of others."

Mahatma Gandhi
(1869 – 1948)

1

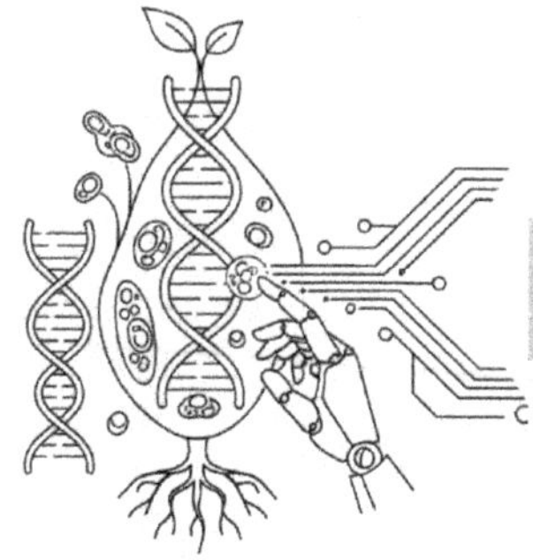

POLLEN ON THE WIND

Every grain of pollen carries a map of survival, drifting like destiny itself — unseen, yet shaping the harvests of tomorrow.

My first childhood memories are from Bhilai in Chhattisgarh, which was a part of Madhya Pradesh in those days. My father (late Shri Mahendra Krishnan) was working in Bhilai Steel Plant. It was a very clean and well-organised city with trees, parks, footpaths, and an efficient flow of movement for people and vehicles. The first child of my parents, I was named Sanjay. By the time I was old enough to start school, my mother (Smt. Shakuntala Devi) had already taught me the basics of the alphabet and daily essentials. She introduced me to the exciting world of riddles, which I now realise were the lessons on lateral thinking. Her common-sense skills were brilliant, which I learned watching her in daily life. My father used to take me for a walk after dinner, and reciting multiplication tables (*Pahada*) while walking used to be fun. This was followed by bedtime stories that continued till I was in class nine. It was during my admission time in school that I came to know that I was not a native of this place, and my ancestors were from Western Uttar Pradesh, a village called Chhur, near Sardhana in Meerut district. Whether it was the preschooling by my parents, the school environment, or whatever, I found school very comfortable, and it was more enjoyable than work.

Once, my father went away for six months to Russia. As I would later know, the Bhilai Steel Plant, located in what is now Chhattisgarh (then part of Madhya Pradesh), India, was established in 1955 with assistance from the Soviet Union. It is a significant integrated steel plant and a major supplier of steel products, particularly long rails for the Indian Railways. The plant was undergoing expansions, including the commissioning of a Wire Rod Mill in those days and the completion of a 2.5 MT stage, and my father was deputed to Russia in that connection. About three years after my father's return to India, I arrived on the scene, and my sister Rina, five years younger than me, also joined the family. My brother Sandeep was born in February 1972, and we had our complete family. A camera and a record player that my father brought earlier from Russia had excellent quality and workmanship, and were our prized possessions.

When I was in fifth grade, my father got a job at the newly established Modi Rubber factory north of Meerut City, in collaboration with Germany. My father was considered an expert in boilers, the heart of any industrial plant, and only a select few in the country had his level of skill and experience. He was offered triple his government salary, and he considered it irresistible. However, later he would know that even if we had asked for ten times, he would have got it, such was the value of his expertise. But whatever, we stayed in a big house in Modi Nagar, the epicentre of Modi Industrial House, on the other side of Meerut, till the Rubber Factory township became a liveable place. I joined a primary school in Modinagar to finish my fifth grade and was introduced to writing on *Takhti* (a portable wooden writing board) using a pen made from reed. Coming from Bhilai, I was shocked since the syllabus was entirely different from what I had studied before. But, because of the tireless efforts of my class teacher, Rani, I was able to cover the entire syllabus, and in the board examination in class 5, I stood 2nd in the whole zone. I realise now that supportiveness and compassion, as exhibited by my class teacher, can uplift the entire fabric of society.

I was admitted to SSD (Shri Sanatan Dharm) Boys' Inter College in class 6 in Lalkurti, Meerut, and by that time, we were allotted a residence at Modipuram. A bus would take us from there to the school. Lalkurti was on the cusp of Meerut Cantonment and the new part of Meerut City called Begum Bridge. Actually, there was no bridge there, and nobody knew the identity of the Begum, or who it was supposed to honour. It was a cluster of upscale markets. Most schools, restaurants and more are in close proximity to the place. Lalkurti, "Red Shirt" in Hindi, must have got its name from the colour of the British soldiers' uniform, who would have been stationed there in the past. The school was good and the teachers were very friendly.

My schooling was disrupted a third time when my father moved to Sindri in Bihar (now Jharkhand) to work for the Fertiliser Corporation of India in a senior position. Sindri is an industrial semi-urban township located in the Dhanbad district. Bihar then had a Higher Secondary School system, where a State-Level examination took place in Class XI instead of Class X, as in Uttar Pradesh. Although based on my performance in the interview, the Headmaster gave me a one-class promotion, from seventh to ninth grade directly, but it exposed me to new subjects like Physics and Chemistry without proper acclimatisation.

My father's elder brother (late Shri Om Prakash), who worked for the railways as Deputy Chief Controller, had 4 of his sons as engineers. It was considered a *fait accompli* that I would become an engineer, but I was aware of my different plans. Although I couldn't fully reason it out, I had an inner sense of studying life forms rather than materials and machines. My father did not put any pressure on me either. And my mother and sister were always my cheerleaders.

The academic environment in Sindri in those days was not good. My father opted for a relocation for me, and I arrived in Lucknow in 1978, while he continued in Sindri. I was admitted to Lucknow Christian College. It was kind of coming to a different country. The British building of the college, constructed in a European style, with high ceilings and long corridors, and curated gardens, was surreal. We could walk to the majestic Lucknow Residency nearby in a few minutes, and the Gomati River flows by it. It was my first exposure to the royal way of life and the cultural mannerisms. Whatever good things you may hear about Lucknow, nothing like living here.

After my intermediate, I wanted to go to Medical School, but much to my dismay, I faced a 'domicile' issue. Since there was no birth certificate confirming my birth in Chhur, I cannot be called from Uttar Pradesh. The next identity, that is, High Secondary, was from Bihar, so I am ineligible to take the CPMT (Combined Pre-Medical Test). There was no other way but to continue at the Lucknow Christian College and graduate in Biological Sciences, which meant studying Botany, Zoology, and Chemistry. I accepted it in the spirit of a Stoic, as they say.

I became a regular visitor to the National Botanic Gardens (NBG), which was rechristened as the National Botanical Research Institute in the same year I arrived in Lucknow. It was a little heaven on earth, as I felt. I was told that it was established in 1800 by Nawab Saadat Ali Khan. Later, Nawab Wajid Ali Shah, the last Nawab of Oudh, dedicated it to one of his favourite queens, Sikander Mahal Begum, and christened this garden as "Sikandar Bagh". During British rule, it became an active centre of horticultural activities and famous British horticulturists, namely, Drs J. Cameron, E. Bonavia, and G.M. Ridley, lived here for their research. Plant scientists Prof. Birbal Sahni and Dr S.K. Mukherjee transformed this horticulture garden into a botanic garden. After independence, it was called 'National Botanic Gardens' and in 1953, the Council of Scientific and Industrial Research adopted it.

I used to spend almost the entire holiday there. I experientially learned to unlock the mysteries of life through the study of plants, the primary bedrock of life on Earth. Plants are the original architects of the planet's breathable atmosphere and the unwavering sustainers of the global food web. Through photosynthesis, plants convert sunlight, water, and carbon dioxide into sugars and oxygen—a process both elegantly simple and immeasurably profound. This conversion is the first link in the chain of life. Every animal, fungus, or microbe that consumes organic matter owes its existence, directly or indirectly, to the photosynthetic powers of plants.

Botany, therefore, is not merely the study of greenery, but of the very engines that drive the biosphere. By decoding the secrets of plant physiology, structure, and evolution, botanists have laid the groundwork for nearly every other field of biology. Botanical gardens served both as scientific laboratories and as living libraries, hosting an ever-increasing diversity of specimens from around the world. This period saw the birth of systematic classification, led by Swedish physician Carl Linnaeus (1707–1778). The binomial naming system—still the backbone of taxonomy—emerged first from the need to organise the bewildering array of plants brought to Europe from newly explored territories.

English naturalist Charles Darwin (1809–1882), best known for his work on animal evolution, devoted much of his later career to studying plants. His experiments with orchid pollination, climbing plants, and carnivorous species demonstrated the subtle interplay of adaptation, competition, and mutualism. The cell, as a concept, was first discovered in plant tissue. The mechanisms of heredity and genetics, too, found some of their earliest and clearest demonstrations in the breeding of pea plants (*Pisum sativum* L.) by Austrian monk-scientist Gregor Mendel (1822 -1884). By the time I completed my graduation in 1982, I was a practising botanist and enchanted with plants; I felt as if the world of agriculture was luring me.

The Govind Ballabh Pant University of Agriculture and Technology (GBPUA&T), Pantnagar is in the Tarai, a vast lowland region situated at the foothills of the south of the Shivalik Hills, the outermost Himalayan range, and characterised by its marshy grasslands, dense forests, and fertile plains. This area acts as a transition zone between the Indo-Gangetic plains and the mountainous Himalayan range, extending across northern India and southern Nepal. Established in 1960, it was the first agricultural university in India. It was initially called Uttar Pradesh Agricultural University before being renamed in honour of Govind Ballabh Pant. The university played a significant role in the Green Revolution, particularly through the testing and propagation of Mexican wheat (*Triticum aestivum* L.) varieties developed by Norman Borlaug.

Pantnagar University was renowned for its "American Way of Education." It was established as a land-grant university in collaboration with the University of Illinois. In its early days, Illinois faculty played a key role in designing the education system and establishing an effective research and extension framework. Typically, six to eight Illinois faculty members would stay in Pantnagar at a time, serving four-year terms over a span of twelve years. The Americans departed from Pantnagar only in 1972, when President Richard Nixon ordered Americans out of the Near East. However, the university's influence and aura persisted. I had a friend named Lokesh Palni. We obtained application forms for both of us. His motivation was that a degree in agriculture could lead to opportunities in the banking sector, which was a popular trend at the time. Both of us received admission.

In the meantime, my father had moved to Talcher in Orissa (now known as Odisha). He seemed to have a great job waiting for him wherever he went, I felt. After finishing my graduation, I was in Talcher when I received intimation of my admission to Pantnagar. I travelled from Talcher to Pantnagar, and from Lucknow, I took the only available narrow-gauge train that made a detour to Lakhimpur-Kheri and Gola Gokharan Nath. After being admitted to the B.Sc. (Agriculture) program, I was allocated accommodation in Nehru Bhavan, a hostel reserved for newcomers and those admitted to the Bachelor of Veterinary Sciences (B.V.Sc.) program. I was shocked to find myself two years older than everyone else in the class and in the hostel, as they all had come after passing their intermediate examination. All new students in Pantnagar are assigned an Advisor from the Faculty who will function as a local guardian. My advisor, Dr S.K. Verma, quickly analysed my predicament and told me to apply for the master's program, where all India competitive examinations were being reconducted. This was unusual, since this exam is normally held only once, but this time it was conducted again, perhaps to shape my destiny.

I got admission to the MSc, to my great relief, as if providence was correcting a man-made mistake. I faced three options: plant physiology, soil science, and genetics and plant breeding. Several of my teachers counselled me that since I got into the master's program with B.Sc. (Botany, Zoology, Chemistry) and not B.Sc. (Agriculture), as I progress in my career, it will be "flagged" by competing candidates, and I may even be hounded by some zealous puritans in selection matters. Therefore, they advised me to pursue plant physiology, as it was taught at the College of Basic Sciences and Humanities (CBSH). I stayed back at the old-fashioned Nehru Hostel instead of moving to the swanky PG Hostel, named Chittranjan Bhawan. From the back side gate, it was a 10-minute walk to the college. Palni had joined Biochemistry. We almost lived together.

Since I was now a master's student, my warden, Dr A.N. Tiwari in the Nehru Hostel, made me the Wing Counsellor, who is entitled to a single occupancy stay and free of charge for performing monitoring services on other students living in that wing, a dozen spacious rooms with double occupancy, especially their nighttime attendance. There was a dosa shanty kind of shop outside the hostel, perhaps the only one in the vast campus, and there would be a queue there as if it were given free. There were also tea and *paan* shops. For local transport, there were shabbily built rickshaws, which were mostly operated by migrants from the eastern part of Uttar Pradesh.

Dr R.C. Pant was the head of the department, and his quota for the students was already full. I was assigned to Dr R.K. Gupta, a UK-educated Punjabi who was comfortable with rustic people like me. I had a brilliant colleague in Taraknath Chakravarty, Ajay and

Minati Pani. D. Subrahmanyam was the only PhD scholar there. We used to work sixteen hours a day, overlapping each other's lunch breaks so that experiments are never disrupted, and Dr Gupta had to tell us to take off at times, manning the work himself. By the time I published two international papers, many of our batchmates were struggling with their first one. After his master's degree, Taraknath Chakravarty went to Jawaharlal Nehru University (JNU), New Delhi, for his PhD and then moved to the USA. Sadly, he left this world too soon. Ajay and D. Subrahmanyam did their PhD from the same department and later joined Indian Council of Agricultural Research (ICAR) as scientists. Minati married Madhu Vrat Singh and moved to the USA.

I received invaluable assistance from Dr Narendra Singh, Professor of Chemistry, specialising in instrumentation. He demonstrated exceptional skill in developing measurement schemes for capturing complex phenomena. Dr D.P. Mishra, Professor of Biochemistry, has a talent for describing processes with the flair of a cricket commentator, making intricate details engaging and accessible. Dr V.S. Rathore, a Plant Physiologist, headed the Radioisotope Facility, which was always in high demand. Radioisotope techniques, particularly separation of radiolabelled metabolites by paper chromatography and exposure to instruments like Planchet Beta Counters, Geiger–Müller Counters, and Liquid Scintillation Counters, have proven to be essential tools in plant physiology, especially for studying water stress in plants.

Pantnagar University had a great library. There was a separate enclosure for the PG students to carry their heavily bound journals and make their reference cards undisturbed by the crowd. The entire campus was dotted with trees, especially the *Eucalyptus* trees (*Eucalyptus globulus* Labill.). Known for its rapid growth and adaptability, they were brought in by the British colonial administration for the pulp industry. It can grow 6-12 feet per year. However, the most significant benefit was that *Eucalyptus* plantations can improve soil health by stabilising the soil, increasing organic matter, and enhancing nutrient cycling. Of course, over time, a counter perspective evolved that growing non-native *Eucalyptus* can be harmful, inhibiting other plant growth through allelopathy and soil hardening, and negatively impacting local biodiversity and soil fertility.

The next step was a PhD. In a scientific career, an MSc is like a Trishanku; to the normal world, it seems like the end has been reached, but in the higher spheres, it is not considered complete and not attainable yet, and you are still midway in your academic journey. My *viva voce* examination delayed the completion of my master's to February, due to a range of reasons, including the postal delays. My thesis examiner, Dr V.K. Kochhar, a scientist at CSIR–National Botanical Research Institute (CSIR-NBRI), Lucknow, conducted the examination in a cordial manner, and I could only join elsewhere in July. One of my friends, Taraknath, was already admitted to JNU, and I thought, first let me meet him, and some lead would come from there. So, I boarded a night bus and landed in Delhi.

There are three great institutions in Delhi – Delhi University (DU), JNU, and ICAR-Indian Agricultural Research Institute (ICAR-IARI). But only the very best could gain entry there. I don't know if it was my attitude or hubris, but I felt I had the best guide to pursue a PhD under. Dr S.C. Maheswari at DU was a legend. But he would not even meet you unless his wife, Dr Nirmala Maheshwari, cleared. She grilled me for an hour but allowed me to meet Dr Maheshwari. He was like a king in his well-furnished room. Even in the daytime, he had lovely curtains on the windows and had an expansive lamp on his table. He spoke to me for a few minutes and asked me to give a seminar on my MSc thesis topic at 4 p.m., which means after a few hours. I thought that was the way Delhi works quickly and swiftly.

Dr Maheshwari sat through my seminar as if meditating with open eyes. After I finished, there was a good discussion, then, after a pause, he said, "You must have known that DU does not provide hostel facilities. So must make your own boarding and lodging." Of course, I did not know that. How have I known also?

The next day, I had an interview for a project with Prof. Sudhir K. Sopory in the School of Life Sciences at JNU. My friend Taraknath was responsible for my first introduction to Prof. Sopory. He is renowned for his pioneering research in plant NextGen Biology and is a very positive person—more interested in what I could do than in what I knew. He immediately encouraged me to join and even wrote a note for hostel accommodation. Being at JNU, I felt as if I was breathing a different air. Unlike Delhi University, where old power structures held sway, JNU was marked by a spirit of freedom and defiance. For a talented young student, the past seemed irrelevant here. After all, India is a country of the poor, and that fact is not only acknowledged but also respected here. That evening, sitting with my friend at a tea stall under a tree, I realised what it truly means to live in a democratic country.

The next day, I arrived at his lab with Taraknath. I met four wonderful people there –Rina Das, Meena R. Chandok and Ashutosh Doshi. They made me feel as if I was at home and one amongst them. In time, when Arun K. Sharma returned from abroad, I had the pleasure of meeting him too. Full of gratitude for Prof. Sopory, who not only accepted me as his student but also provided me with a roof in Delhi, and eager to learn, I felt like a novice in a monastery. In a few weeks' time, I have learned how to separate proteins and prepare a tissue for examination.

As a new student, the JNU campus feels like a vibrant and dynamic haven of learning and diversity. Nestled in the heart of New Delhi, the campus is expansive, with lush green lawns, sprawling trees, and peaceful spots for reflection or discussion. The iconic Sabarmati Hostel and the distinctive, modern administrative buildings stand out against the natural backdrop. Walking through the campus, you can hear lively debates, campus events, and the hum of student activities everywhere. The atmosphere is intellectually stimulating, with a sense of camaraderie and activism that are deeply ingrained in its identity. It's a place where ideas flourish, friendships bloom, and every corner seems to echo with the spirit of inquiry and change.

But when I thought that was it, I got an interview call from ICAR-IARI. The good point was I only had to board a local bus to be there, merely 14 km, and it took not even half an hour. Nestled in the heart of New Delhi, in the forest area called the Ridge. The spacious ICAR-IARI campus stands like an island with a lighthouse, radiating the light of scientific discovery, innovation, and academic excellence across the subcontinent and beyond. For over a century, ICAR-IARI has been synonymous with agricultural advancement in India, serving as both a guardian of tradition and a vanguard of progress. As a premier postgraduate institution, ICAR-IARI offers MSc and PhD degrees across a spectrum of disciplines.

I cracked the interview, and when it came to selecting a guide, I wrote my choice for the legendary Dr Suresh K. Sinha, and I don't know how and why he accepted me. Dr Sinha was an outstanding plant physiologist and a master of stress physiology. He always thought beyond what most of us could imagine, and would open new possibilities in science. Despite his accomplishments, he was a very simple and approachable person, radiating positivity and inspiring everyone around him with his dedication, curiosity, and commitment to excellence. Such was his eminence and renown that even from my MSc days, I had aspired to work under his guidance. I still remember him saying, "Sanjay, ICAR-IARI is more than an institution; it is a living, breathing citadel of knowledge where ideas are born, nurtured, and transformed into reality. Its corridors echo with the footsteps of pioneers, its fields bear witness to generations of toil and triumph, and its spirit continues to inspire all who seek to harness the power of science for the greater good. More than we teach, we open minds and hearts here. Are you ready for the journey ahead?"

I told him firmly, "Sir, I will only go home after completing my PhD." Dr Sinha looked at me with a gentle smile and replied, "But that could be many years down the road, but with dedication and hard work, you will get there." I responded with conviction, "Sir, I will work under your guidance tirelessly until that day comes, whenever it may be. But I am not leaving until I have earned that degree." He seemed a bit surprised by my unwavering resolve, perhaps doubting whether I truly meant what I said. Yet, sensing the sincerity in my words and not wishing to offend or discourage my determination, he simply nodded and said, "Alright, child. Get down to work." That moment cemented my commitment—no matter how long the journey took, I was determined to stay focused and see my goal through.

This is not the place to describe all that I have learnt under the legendary scientist, except confessing that he transformed me. Dr Sinha did not tell me directly, but made me internalise that it takes a lot of effort and research to find a solution. Agricultural research, particularly in the realm of plant science, is inherently a slow and meticulous process. When working with plant species that predate human civilisation itself, we are delving into a realm largely shrouded in mystery. These ancient plants harbour vast amounts of genetic information, much of which remains unexpressed and hidden within their germplasm. Our understanding of this genetic wealth is still limited, and any attempt to hasten the process or seek shortcuts can inadvertently cause us to overlook critical clues embedded within their DNA. Such oversight can lead to costly mistakes when these plants are cultivated in real-world fields, potentially resulting in reduced yields, compromised resilience, or unforeseen vulnerabilities. Therefore, a patient, observant approach is essential—one that involves becoming attuned to the subtle signals the plant offers. By immersing ourselves fully, without rushing or feeling under pressure, and by forging a true connection with the plant, we can discern the right moment for intervention. Only then can we make informed, thoughtful changes that honour the complexity of these ancient genetic repositories, ensuring sustainable and successful agricultural practices.

Back in 1986, I began my journey at ICAR-IARI with nine others in my batch: Prabhjeet Singh (retired as Professor from Guru Nanak Dev University, Amritsar), Deepak Mathur (currently working for a liquid fertilizer company in Australia), Sewa Ram (retired as Principal Scientist from ICAR–Indian Institute of Wheat

& Barley Research, Karnal), Malik Zainul Abdin (currently serving as Professor at Jamia Hamdard University), Vandana Bhargava (did not pursue a profession), Surma Mitra (did not pursue a profession), Renu Kapoor (did not pursue a profession), Manoranjan Kar (retired as Vice Chancellor from Odisha University of Agriculture & Technology, Bhubaneswar), and G.B. Khandagale (retired as Professor from Vasantrao Naik Marathwada Krishi Vidyapeeth, Parbhani).

Our lab at the Water Technology Centre was a place of inspiration and guidance. Dr Renu Khanna Chopra, a gifted scientist, worked passionately on stress physiology. Dr P.K. Aggarwal, who began as a plant physiologist and later became an expert in crop modelling, was supportive of colleagues and remained focused on his work. We also had Kailash C. Bansal, a highly energetic PhD scholar, who later retired as Director of ICAR–National Bureau of Plant Genetic Resources, New Delhi.

I stayed in Room No. 14 at Hemant Hostel, though it served more as a place to rest. Most of my time was spent in the lab or the library. Since I rarely visited home, my parents would occasionally come to see me, bringing with them the comfort of home and strengthening my resolve to continue. These years shaped me in scientific learning, but through the joy of discovery. The people I met left lasting impressions that I still remember; their encouragement and support have remained part of my life since then.

From Dr Sinha's perspective, I completed my PhD in less than four years. However, my *viva voce* would take another year for some administrative reasons. In the meantime, I secured a position at the CSIR-Institute of Himalayan Bioresource Technology (CSIR-IHBT; CSIR Complex Palampur) in Palampur, nestled in the western Himalayas of Himachal Pradesh. Coincidentally, my director, Dr N.K. Jain was travelling on the same train from Delhi for my onboarding. Upon learning this, he arranged for me to disembark at Pathankot Railway Station and travel in his car to Palampur. As we entered the CSIR-Institute of Himalayan Bioresource Technology campus, with the majestic Dhauladhar Mountains as a backdrop, I was utterly awestruck. It felt as though the pollen, carried by the wind for years, had finally found its landing on the stigma of my quest.

•••

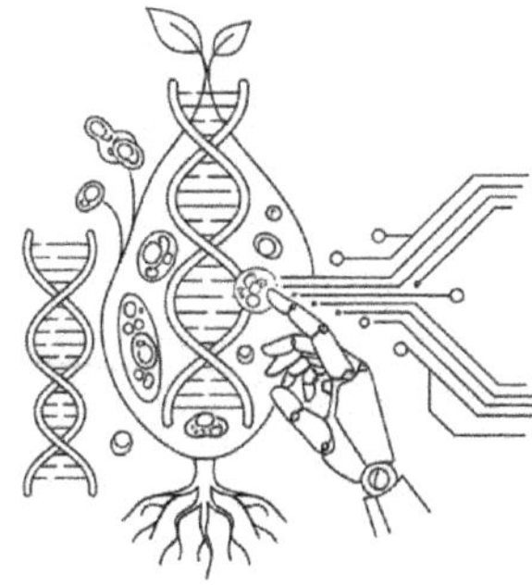

2

THE LURE OF THE NEW

Curiosity is the bridge between what is known and what is possible; it is the ember that turns discovery into destiny.

A new scientific lab is a blank canvas. For many, this clean slate is irresistible. In a space unencumbered by tradition, every protocol, routine, and project is open to creative reimagination. Researchers are often encouraged to voice their ideas, propose novel methodologies, and shape the direction of the lab's research agenda in ways that might be unthinkable within the rigid hierarchy of a longstanding group. The absence of established dogma can foster a spirit of innovation and a willingness to take bold intellectual risks.

When the Kangra district was still a part of Punjab, the State Government considered setting up the National Biological Research Institute (NBRI) at Palampur, as a constituent establishment of the Council of Scientific and Industrial Research (CSIR), an autonomous society under the Ministry of Science and Technology and allotted about 1,550 acres of land at Holta, Palampur. But in an unexpected turn of events, the Kangra district was merged with Himachal Pradesh (HP) and the issue of setting up NBRI lost priority. The CSIR, however, marked its presence in the state in 1973 by creating an Extension Centre of its Regional

Research Laboratory (RRL) Jammu at the Bundla Tea Estate at Palampur. But it would take a decade before the decision to create a CSIR Complex at Palampur was taken and another three years for the inauguration of the first building.

The Palampur area had a thriving Tea Industry dating back to the late nineteenth century, which, for various reasons, had been languishing since then. The local Tea Cooperative Factory, Palampur, conveyed the need for a strong R&D backup to revive the ailing tea industry, paving the way for the revival of 'Kangra Tea Industry' in this part of the region. Dr N.K. Jain, a renowned Tea technologist at Tocklai Tea Research Institute, Assam, was appointed as Coordinating Director to head CSIR Complex Palampur. New labs tend to be nimble. With fewer entrenched habits and administrative bottlenecks, decisions can be made rapidly. The energy is palpable—every small success, from acquiring new equipment to publishing a first paper, is celebrated as a monumental milestone.

Yet, the same freedom and flexibility that fuel creativity in new labs also breed uncertainty. Researchers must contend with the daunting reality of limited resources. Funding may be sporadic, equipment scarce, and institutional support minimal. Unlike established labs that often have extensive stocks of reagents, a steady stream of grants, and well-oiled administrative machinery, new labs may find themselves negotiating for every pipette and every hour of technician support. Dr R.P. Sood joined in May 1985 as the first scientist, followed by Dr P.K. Nagar, Dr Arvind Gulati and Dr Anil Sood as his associates.

The crystal grew fast thereafter. Dr D.N. Chakrabarty, Dr D. Mukherjee, Dr S.D. Ravindranath and Dr L.M.S. Palni had arrived. For most of them it was a gamble. The lab's track record was non-existent. The pressure to produce results and secure funding was high, and it trickled down and created a sense of instability among the new members, including me, who had just arrived.

I was 27 years old at that time. I initially stayed at the guest house for about a week, acclimating myself to the environment.

Soon after, I was allotted a house that had originally been meant for the tea labourers but was later converted into a one-bedroom staff quarter within the campus at Bharmat, though it was located about 2 km away in the farming area. Unlike the lively buzz of Delhi, this place was enveloped in profound silence, offering a stark contrast that felt almost meditative. I could hear the rustling of leaves and the chirping of birds as if nature itself was engaging in a direct communication, creating a peaceful and almost spiritual atmosphere. This serene environment allowed me to disconnect from the chaos of city life and focus deeply on my work, fostering a sense of tranquillity and reflection that was essential for my growth and productivity during that period.

Here again I received most unexpected support. Dr Palni, Head of the Division of Biotechnology and my boss, was the elder brother of my colleague Lokesh Palni in Pantnagar. Born in Almora in 1953, Dr Palni earned a PhD from the University College of Wales, Aberystwyth, in 1980. He then spent almost a decade as a Post-Doc and Research Faculty at the Australian National University, Canberra, Australia. Dr Palni became my mentor and articulated my ideas about plant growth substances into research hypotheses. Established labs have networks built over years, if not decades, which facilitate collaborations, co-authorships, and resource sharing. In a new lab, those networks must be painstakingly constructed. Yet, this process can forge deep bonds among group members and with external collaborators, as everyone is invested in mutual success.

I met Richa, a local graduate trainee at the centre. There is something mystical that resonates when you meet certain people, and I felt that, but I was too shy to express it to her or anyone. She left soon for her master's degree in economics at Dharamshala, about 40 km northwest of Palampur, the administrative headquarters of the Kangra district and the biggest town in the region. Her father (Dr Surinder K. Mishra) was the professor and Head of the Department of Horticulture and Forestry at Himachal Pradesh Krishi Vishvavidyalaya (HPKV; rechristened as Chaudhary Sarwan

Kumar Himachal Pradesh Krishi Vishvavidyalaya (CSK HPKV) in 2001), at Palampur, and was all that I could know. Anil Sood and Arvind Gulati, my seniors at work and well conversant with the town, used to take me out. There were several educational institutions, including a regional centre of the ICAR-Indian Veterinary Research Institute (ICAR-IVRI) and a unit of the ICAR-Indian Grassland and Fodder Research Institute (ICAR-IGFRI). No wonder that in the NCERT textbook of class 9 economics, there is a chapter "The Story of Village Palampur" to illustrate various aspects of rural life, including farming, non-farming activities, and the factors of production.

I was appointed to the position of a flower physiologist. The work of a flower physiologist involves exploring the biological and biochemical processes that govern the development, growth, and function of flowers. Scientific aspects include studying respiration, hormone regulation, such as the roles of auxins, gibberellins, cytokinins, and ethylene, which influence flowering, blooming, and senescence. They investigate the signal transduction pathways that control flowering time and floral organ formation, often using NextGen Biology techniques to analyse gene expression patterns. Flower physiologists examine physiological responses to environmental factors like light, temperature, and water availability, understanding how these cues trigger flowering or influence floral morphology. Additionally, they analyse nutrient uptake, transport mechanisms, and cellular activities within floral tissues to comprehend the complex interactions that result in the successful development of flowers, which are vital for plant reproduction and crop productivity.

As my PhD *viva* was pending and the umbilical cord from ICAR-IARI was not severed yet, I would go there every 2-3 months and share with Dr Sinha and other colleagues about my work in Palampur. Thanks to this connectivity, I developed insights into plant electrophysiology, concerning the electrical signals and responses within plant tissues, similar to how nerve signals operate in animals. Plants generate and conduct electrical impulses in response to environmental stimuli such as light, touch, and injury. These electrical signals play a crucial role in coordinating physiological processes like the opening and closing of stomata, the movement of nutrients, and the response to stress. Researchers measure changes in electrical potential and observe how these signals propagate through various plant tissues, providing insights into plant behaviour and communication mechanisms.

Jagadish Chandra Bose (1858–1937) was a pioneer in this field. He was one of the first scientists to demonstrate that plants exhibit electrical responses and can perceive external stimuli. Bose developed innovative techniques to detect electrical signals in plants and proved that they have a form of sensitivity and responsiveness akin to animal nervous systems. His experiments showed that plants respond to touch, light, and other stimuli with electrical impulses, challenging the notion that plants are passive organisms. Bose's work laid the foundation for understanding plant electrophysiology, establishing plants as responsive, living systems capable of complex communication and adaptation.

Guided by Dr Sinha, I hypothesised that drought impacts a plant's electrophysiology by altering its electrical signalling and responses. During drought stress, plants experience changes in ion fluxes, membrane potential, and electrical conductivity, especially in tissues like the roots, leaves, and vascular system. These changes are part of the plant's physiological response to conserve water and initiate defence mechanisms. For example, drought conditions often cause an increase in the electrical potential difference across cell membranes, which triggers the closure of stomata to reduce water loss. Additionally, drought-induced electrical signals can propagate through the plant, activating stress-responsive pathways in distant tissues. These electrical responses help the plant initiate molecular and hormonal changes critical for survival under water-deficient conditions, including the production of stress hormones like abscisic acid, which further modulates cellular activity and water conservation strategies.

Instrumentation for plant electrophysiology spans a wide spectrum, from basic tools to advanced systems designed for detailed analysis. At the most fundamental level, glass microelectrodes are used to measure membrane potentials and ion fluxes within plant cells. These fine-tipped electrodes can penetrate cell walls and membranes to record electrical activity directly from individual cells. Surface electrodes are also employed, attached to external surfaces like leaves or roots, to capture electrical responses without invasive procedures. Galvanometers, highly sensitive devices, are used to detect and measure tiny electrical currents or potential difference generated in plant tissues, providing initial insights into the electrical behaviour of plants.

As scientific techniques advance, intermediate instruments such as oscilloscopes become essential. Oscilloscopes display electrical signals in real time, allowing researchers to analyse the amplitude, frequency, and waveform of electrical impulses in response to various stimuli. Data loggers further supplement this analysis by recording electrical signals continuously over extended periods, enabling the study of dynamic physiological responses under different environmental conditions.

For more sophisticated studies, advanced instruments are employed. Electrophysiological amplifiers with high gain and noise reduction capabilities are used to detect minute electrical signals with clarity. Microelectrode arrays allow the simultaneous recording of electrical activity from multiple points within a plant, facilitating spatial mapping of electrical responses across tissues or organs.

Additionally, signal processing systems equipped with specialised software can filter, analyse, and interpret complex electrical data, providing deeper insights into plant signalling pathways. Emerging imaging techniques, such as the use of voltage-sensitive dyes combined with microscopy, enable visualisation of electrical activity at cellular or tissue levels *in vivo*. Collectively, these instruments provide powerful tools for understanding the electrical mechanisms underlying plant responses to their environment, physiological regulation, and internal communication. I discussed with Dr Palni how to do this work. There were limited facilities, and even for furniture, there was one round stool in the lab, which I would share with Dr Anil Sood —a small detail that spoke to the companionship we enjoyed in those days. ICAR-IARI had all these facilities, but while I could discuss my ideas there, I could not carry out the work of Palampur. Eventually, it was K. Balachandra Hebbar who carried forward this line of work under the guidance of Dr Sinha with remarkable dedication. He not only undertook the experimental studies in electrophysiology but also did a wonderful job of completing his PhD on it, setting a benchmark for future work in this area.

By this time, Richa had completed her post-graduation, and when we met again, I proposed to her. Our parents readily agreed. It was a leap year, and we married on the second Sunday of the month, on February 9, 1992. The initial days of married life are typically filled with a blissful sense of wonder and happiness that feels almost heavenly. Every moment seems imbued with a special magic, as if a new, divine energy has begun to flow through you, awakening a deep sense of joy and purpose. The companionship of a loved one brings unparalleled comfort, and each shared experience feels like a beautiful discovery, intensifying the bond and filling the heart with warmth. This phase is characterised by a gentle, uplifting serenity, where love and hope seem to bloom effortlessly, creating a euphoric atmosphere that lifts the spirit and inspires a profound sense of contentment and fulfilment.

In 1992, Dr Palni left to join Govind Ballabh Pant Institute of Himalayan Environment & Development as Head, Division of Environmental Physiology and Biotechnology at Kosi-Katarmal, Almora, Uttarakhand. We had a dinner meeting before he left, wherein Dr Palni motivated me to work abroad and "taste" how science is practised there. For the adventurous scientist, the flip side of working in a new lab is the exhilarating possibility of leaving a mark—not just on a project, but on the very culture and direction of scientific discovery. For others, the wisdom and networks of an established group may provide the best environment for learning and advancement. Both paths offer lessons and rewards; the key is to recognise which suits your ambitions and to embrace, with open eyes, the journey ahead.

My PhD guide, Dr Sinha, also encouraged me to apply to world-class universities for advanced training in plant NextGen Biology, as he felt that scientific skills can only be learnt by working in well-equipped laboratories with people around you showing you how things are actually done. Scientific research is a highly skilled and meticulous field that often begins with initial insights or approximate progress, which may only amount to about 10% of the overall understanding. Beyond this early stage, further advancements require rigorous experimentation, detailed observation, and critical analysis of underlying processes. To achieve meaningful progress, researchers rely on sophisticated instruments and advanced techniques capable of detecting subtle phenomena that are not visible to the naked eye. These tools enable scientists to deconstruct complex biological, chemical, or physical mechanisms, leading to more precise data and deeper insights. The combination of skill, technical expertise, and advanced instrumentation is essential for pushing the boundaries of knowledge and transforming preliminary ideas into comprehensive scientific understanding.

I kept close to my heart the words of Dr Sinha and Dr Palni. The allure of the new is the greatest motivation for a scientist, driving them to explore uncharted territories and challenge existing knowledge. This fascination with discovery fuels curiosity and inspires innovative thinking, encouraging scientists to develop novel hypotheses and experimental approaches. The promise of uncovering groundbreaking insights or revolutionary technologies propels their dedication and perseverance, ultimately pushing the boundaries of human understanding. In essence, the pursuit of the new embodies the spirit of scientific inquiry, making it the most compelling force behind scientific advancement. Our director, Prof. Akshay Kumar Gupta, supported me in every aspect.

In March 1993, I was awarded the BOYSCAST (Better Opportunities for Young Scientists in Chosen Areas of Science & Technology) Fellowship to work at the Lab of Texas Tech University, Lubbock, USA, with Prof. Henry T. Nguyen for one year, which I availed in 1994. The BOYSCAST Fellowship is a prestigious program funded by the Department of Science and Technology (DST) in India. It aims to attract talented young scientists and engineers to pursue advanced research and training abroad in various scientific and technological fields. The fellowship provides financial support for research activities, including travel, accommodation, and research expenses, allowing recipients to enhance their skills and knowledge through international exposure to develop highly skilled researchers who can contribute to the country's scientific and technological progress.

Dr Paramvir Singh Ahuja, a renowned plant breeder, had joined as Head, Plant Biotechnology and Biodiversity Conservation Division. He had worked at McGill University in Montreal, Canada, Ohio State University in Columbus, USA, Plant Breeding Institute, Cambridge, UK and Nuclear Research Institute at Jülich in Germany and created a world-class protoplast fusion facility at CSIR-Central Institute of Medicinal and Aromatic Plants (CSIR-CIMAP), Lucknow. Dr Ahuja asked me to be clear about my work after my return and asked about where my heart was. I said I was deeply interested in the protein stability in plants. How can certain protein molecules, such as thaumatin, be stabilised so as not to be affected by boiling or extreme cold later? A sagacious person, Dr Ahuja blessed me and said that this is a challenging area and can make a molecule like thaumatin a real alternative to artificial sugar someday. Later, when Dr Ahuja went to Brazil, he brought back with him the seeds of stevia [*Stevia rebaudiana* (Bertoni)]. He could not, however, obtain *Thaumatococcus daniellii* (Benn.) Benth (source of thaumatin protein), and that was how the CSIR complex eventually began working on sweeteners other than sugar. Later, in March 2018, another low-calorie natural sweetener plant, monk fruit [*Siraitia grosvenorii* (Swingle)], was introduced by Dr Probir

Kumar Pal and the first fruits were harvested in November 2018. This herbaceous, perennial plant is native to the southern parts of China. The extract from the processed monk fruit is 200-300 times sweeter than sucrose, due to the presence of triterpene glycosides, mogrosides.

I also discussed with Dr Ahuja the idea of deciphering the molecular mechanism of winter dormancy in tea. This is a phenomenon where the growth of the tea plant slows down, and the production of the apical bud along with the two accompanying leaves, commonly known as the "two leaves and a bud," comes to a stop. These are the very parts used to make quality tea, and when dormancy sets in during the winter season, it means no harvest for the growers and no work for the labourers. I still remember how readily this idea was supported by the erstwhile Director, Dr Gupta, and by Dr Ahuja, which gave me the encouragement to take it forward.

Also, Dr Ahuja asked me if I would like to work on high altitude plantations. In his view, deciphering how plants survive the cold deserts was the logical culmination of my work in protein stabilisation in plants.

Richa was on her family way, but she encouraged me to move on to Texas Tech University as such opportunities do not come often for middle-class people like us. Then her parents were around, which gave me confidence. I took an Air India flight on March 20, 1994 from New Delhi to New York via London Heathrow. From New York, I flew by Delta Airlines to Dallas, and then an hour flight to Lubbock in Texas. I could see the terrain between Dallas and Lubbock transitioning from relatively flat to gently rolling hills and felt a strange feeling as if I was flying over Himachal Pradesh's lower hills, also known as the Shivalik Hills, which are located in the southern part of the state and include parts of Kangra, Hamirpur, Una, Bilaspur, and the lower regions of Mandi, Solan, and Sirmaur districts.

Texas Tech University, located in Lubbock, Texas, is a renowned public research university known for its strong academic programs, innovative research, and vibrant campus community. Established in 1923, it offers a wide range of undergraduate, graduate, and professional degrees across various fields, including engineering, business, health sciences, and arts. The university is also home to notable research initiatives and athletic programs, particularly its successful football team. With a focus on fostering experiential learning and community engagement, Texas Tech University plays a significant role in higher education and economic development in West Texas. Its expansive campus and diverse student body contribute to a dynamic and inclusive environment for students and faculty alike.

I worked in the laboratory of Dr Henry T. Nguyen, a professor of genetics at Texas Tech University. A PhD from the University of Missouri (MU) in Columbia, Dr Nguyen is considered the world's leading scientist in developing genomics-assisted breeding technologies to improve crop yield, abiotic stress tolerance, and nutritional value. He would serve as the Founding Director of the National Centre for Soybean Biotechnology (NCSB) established by the United States Congress in 2004.

Prof. Bin Wang from China and Erin, an American, were also working in Dr Nguyen's Lab. Prof. Wang had rented accommodation in Lubbock, and by the time I arrived, he had to return to China. He offered that place for me to stay. In my one year there, he appeared only once or twice for a few days. So, I was living in Prof. Wang's accommodation without Prof. Wang. Whenever he came, we shared the rent; otherwise, I managed it alone. Erin, our lab manager, was efficient and thoughtful. She kept the lab running nicely.

It was like living in another world—strange, distant, yet softened by the kindness of people who became family. Chandrashekhar P. Joshi, known simply as Shekhar, with his wife Rama and their daughter Apoorva; Oswald Crasta and his wife Jacintha, with their daughters Ranjitha and Natasha; and Dr R. Chandrababu, newly arrived from Coimbatore, with his wife Umadevi, son Ramesh

Chander, and daughter Sindhujaa—all of them wove a circle of warmth around me, a small home away from home.

When my daughter was born in Palampur on September 24, the news reached me only two days later. Such was the state of the telephone lines then—Palanpur and Palampur confused at the operator's desk, voices misdirected across the wires. Yet the delay could not dim the joy. These families, sharing my happiness, gathered to celebrate the arrival of my first child, whom I would not see until six months later. We named her Avisha—a Sanskrit name, meaning "one blessed by the grace of God." In that name, I felt both the distance and the nearness of grace, carrying me through those long months of waiting.

My exposure to Dr Nguyen's laboratory helped us to establish NextGen Biology facilities at Palampur. One important step was creating a facility to examine superoxide dismutase (SOD), an enzyme that facilitates the conversion of superoxide radicals into hydrogen peroxide, which then requires regulation to prevent cellular damage. Hydrogen peroxide (H_2O_2) metabolism in plants plays a dual role as both a signalling molecule and a potential source of oxidative damage. It is primarily generated as a by-product of various metabolic activities, such as photosynthesis in chloroplasts and respiration in mitochondria. Other enzymatic sources of H_2O_2 include peroxidases, amine oxidases, and NADPH oxidases, which actively produce H_2O_2 in response to developmental cues or environmental stimuli.

To maintain cellular homeostasis, plants possess efficient scavenging systems that detoxify excess hydrogen peroxide. Key enzymes involved in this process include ascorbate peroxidase (APX), which reduces H_2O_2 to water using ascorbate as an electron donor, and catalase, which decomposes H_2O_2 directly into water and oxygen, mostly in peroxisomes. Glutathione peroxidases also contribute to detoxification by reducing H_2O_2 with glutathione. This balance between production and scavenging of H_2O_2 is crucial for allowing it to serve its role as a signalling molecule without causing oxidative stress that could damage cellular components.

Besides, H_2O_2 acts as a secondary messenger in various plant responses to environmental stresses such as drought, salinity, pathogen attack, and wounding. Its production in specific cellular locations and regulated scavenging enable plants to perceive and adapt to changing conditions efficiently. The antioxidant system, which includes molecules like ascorbate, glutathione, and tocopherols, works in tandem with enzymatic pathways to finely tune the levels of H_2O_2. Overall, proper regulation of hydrogen peroxide metabolism is essential for plant development, stress resilience, and maintaining cellular health.

CSIR Complex Palampur was renamed as CSIR-Institute of Himalayan Bioresource Technology in January 1997. In April 1998, when Dr Ahuja took over as the Director, he took me to the Golden Temple. It was a transformative experience. I am a theist but not a religious person. You see God not necessarily as an individual creator who designed the universe or as a monitor overseeing human affairs, but rather as the fundamental essence of life itself—an ongoing process or principle that underpins existence.

This view aligns with a more abstract or pantheistic understanding of divinity, where God is equated with the fabric of reality, the flow of life, or the natural order. It emphasises an intrinsic connection to the universe, viewing God as an integral part of the ongoing evolutionary and energetic processes that sustain and evolve life. In this sense, your belief might focus on experiencing the divine through an awareness of the interconnectedness and dynamism of all living things, rather than through traditional doctrines, rituals, or dogmas. It's a perspective that invites personal reflection and an appreciation of the divine in the natural, ever-changing fabric of life itself.

On that trip, Dr Ahuja explained to me that light symbolises the divine presence as pure, illuminating, and ever-pervading. It represents the universal force of life, knowledge, and spiritual awakening, dispelling ignorance and guiding in the darkness of doubt. Just as light sustains growth and reveals truth in the physical world, seeing it as a divine essence highlights qualities of purity, goodness, and omnipresence. This perspective invites us to recognise the divine not as a distant creator but as the luminous energy that flows through all of existence, inspiring clarity, connection, and enlightenment in our journey through life.

Our first research trip to Kaza in Spiti Valley was in 1995, following my return from the USA. We used a four-wheel drive vehicle, the Maruti Gypsy, and another car followed with equipment and provisions. We drove all the way up to Rohtang Pass in the Pir Panjal Range of the Himalayas in Himachal Pradesh. Situated at 13,000 feet, it connects the Kullu Valley with the Lahaul and Spiti Valleys. After Marhi, there was no proper road, and we had a feeling of surviving by the hour when traversing the makeshift roads. Then, after reaching Rohtang Pass, we had a contrasting view, a lush green valley to our south and a barren desert towards our north.

We reached Kunzum Pass, via Koksar and Chhatru, at 15,000 feet, which connects the Lahaul and Spiti Valleys. There are 15 hairpin turns here, offering breathtaking views of the surrounding mountains, glaciers, and valleys. I struggled to breathe while walking. I wondered how plants would breathe here. Proceeding further, we reached Kaza, the subdivisional headquarters of the remote Spiti Valley, at 12,000 feet. Overlooked by high mountain ridges on the northeast and southwest sides, the Spiti River runs eastward past Kaza. But for the river, the valley is arid and rugged. On the way to Kunzum Pass from Koksar, finding not even a blade of grass anywhere, I had a feeling as if I had arrived on some other planet.

Grass and other vegetation are unable to grow in cold deserts primarily due to the extreme environmental conditions. One of the main reasons is the very low temperatures, especially during winter, which inhibit seed germination and the metabolic processes necessary for plant growth. Most plants require a certain temperature range to carry out photosynthesis and nutrient absorption, and these conditions are rarely met in cold deserts.

Another significant factor is the scarcity of liquid water. Although snow and ice are abundant, the surface water often remains frozen due to permafrost, making it inaccessible for plants. Since water is essential for photosynthesis and nutrient uptake, its absence in liquid form prevents most plants from establishing themselves. Additionally, the soils in cold deserts are typically nutrient-poor because the slow decomposition of organic matter in cold temperatures limits the recycling of nutrients, further hindering plant growth.

The short growing seasons also play a crucial role. In these regions, the window of suitable conditions—when temperatures are warm enough for growth—is very brief. During this limited period, plants must complete their life cycle, which is a challenge given the harsh environment. Moreover, high levels of solar radiation and strong winds increase evaporation rates, causing further water loss and creating an even more hostile environment for vegetation.

Finally, the presence of permafrost layers beneath the soil prevents roots from penetrating deeply and accessing water and nutrients. Only specially adapted plants like lichens and mosses can survive in these conditions, while grasses and more common plants typically lack the necessary adaptations. In summary, the combination of severe cold, lack of accessible water, poor soil nutrients, and limited growth periods makes it impossible for most grass species to thrive in cold deserts, leaving only hardy, specialised plants able to survive there.

Our second child, a son, arrived on March 28, 1999. We named him Vayun. The name Vayun was taken from the Yajurveda Richa.[1] This hymn is part of a series of praises dedicated to Agni, the fire god, who holds a pivotal position in Vedic rituals and spirituality. Agni is seen as an archetype of the priestly function—sacred, pure, and divine. As 'the holy,' he embodies spiritual purity and is revered as a divine presence within life. He is venerated for being wise

---

1 अग्ने नय सुपथा राये अस्मान्विश्वानि देव वयुनानि विद्वान्। युयोध्यस्मज्जुहुराणमेनो भूयिष्ठां ते नमउक्तिं विधेम ॥ (Yajurveda, Chapter 40, Verse 16)

and swift. 'Wise' indicates his knowledge and understanding of divine and earthly matters, while 'swift' signifies his quick action, conveying divine messages and blessings rapidly. We named our son with the hope that he would be swift in implementing wise actions.

There was a deep sense of parental satisfaction. Professionally, I focused on high-altitude environments to investigate the unique adaptations of plants living in these regions, combining fieldwork and laboratory research to understand their physiology and genetic makeup. This knowledge helps in fostering sustainable plant growth in extreme environments and addressing ecological challenges. What is science but an eternal quest, driven by the irresistible lure of the new, forever seeking to unveil the mysteries of the universe and expand the horizons of human knowledge?

•••

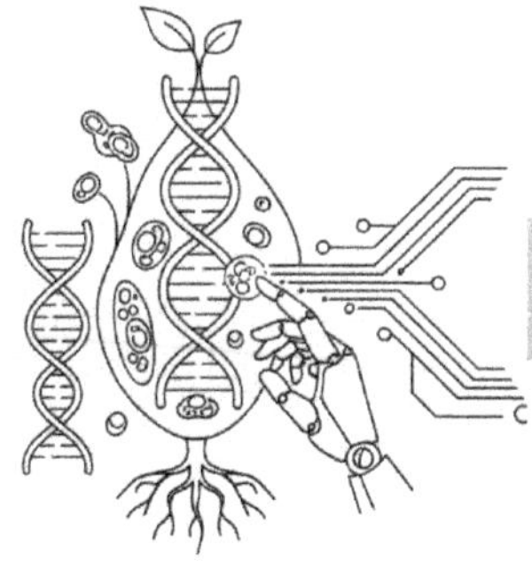

3

DECODING LIFE

To read the language of genes is to glimpse the script of creation — not to conquer it, but to serve life more wisely.

Lahaul-Spiti is remote but culturally rich. Historically, it has been part of the trade routes connecting Tibet, Central Asia, and the Indian plains, where goods such as salt, wool, and grains were bartered. The area has been influenced by Tibetan culture, Buddhism, and various indigenous traditions for centuries. A part of the larger Trans-Himalaya region, Lahaul is a lush terrain that was historically governed by local rulers and chieftains. Spiti, meaning "the middle land," served as a key strategic and spiritual centre. Kye Monastery, called Kye Gompa, is the largest monastery in the valley and is situated near Kaza, on a hilltop at an altitude of 4,166 meters. Over 300 monks reside there.

It was not easy to travel. Even though the car was not available for junior scientists and staff, we would take local buses, changing at different towns and putting our equipment on the bus roof. Also, we had to take provisions for about a month's stay. From Palampur to Kaza, the primary route was through Manali, Marhi, Rohtang Pass, Koksar, Chhatru, Batal, Kunzum Pass, and Losar. Batal, an important village between Chhatru and Kunzum Pass, was a key stop for its basic facilities.

When Rohtang Pass was closed due to snow, specifically from November to April, we would go via Shimla, Narkanda, Rampur Bushahr, and across the Wangtu Bridge over the Sutlej River, to Kinnaur and Spiti. The route continues via Reckong Peo and Tabo, to Kaza. We also used to take a third route that diverges from Kullu, crosses Jalori Pass, and joins Rampur Bushahr, merging with the Hindustan-Tibet Road to pass the Wangtu Bridge, Reckong Peo, and Tabo before arriving at Kaza. There is a 1,000-year-old monastery at Tabo, known as Tabo Chos-Khor Monastery. Renowned for its ancient murals and scriptures, it is called the "Ajanta of the Himalayas."

Research on high-altitude plants holds significant potential for uncovering secrets that could advance our understanding of life and its adaptations. These plants have evolved extraordinary mechanisms to survive extreme conditions, such as intense UV radiation, freezing temperatures, limited water, and nutrient scarcity. By studying their unique biochemical pathways, protective pigments, and stress-resistance strategies, we could gain insights into how life persists under environmental stressors that would be lethal to many other organisms.

Dr Ahuja, like some ancient Rishi, used to motivate us to take this arduous journey as a mission of decoding life. Plants are some of the earliest organisms on Earth, existing long before humans. They produce a wide variety of molecules, such as chlorophyll, alkaloids, and secondary metabolites, which have been invaluable in scientific research. These molecules can help us understand fundamental biological processes and the origins of life, as they often contain ancient biochemical pathways. Studying plant molecules can provide insights into the evolution of life and may even reveal clues about how complex organisms, including humans, came to be on Earth, Dr Ahuja used to say.

I used to feel as Plato would have felt hearing Socrates. Our salaries were meagre. Working conditions were harsh, and except for this zeal for discovering secrets of life in plants, there was no rhyme or reason in returning every quarter after a few weeks in

Palampur to unload the plant material for further analysis and replenish laboratory materials. There were no telephones, and for weeks, we had no contact with our families. They would not even know if we encountered an emergency in the field.

My wife, Richa, raised our children in a most courageous manner. She had a profound respect for our work and never complained or shared her hardships with me. Once, when I was sharing with her how plant biology plays a crucial role in understanding adaptation and disease pathways and the sophisticated adaptive systems plants evolve to defend against pathogens, involving mechanisms like tolerating harsh environmental conditions, pathogen recognition, signalling pathways, and production of defence compounds, she smiled and said that plants are fortunate not to move around and have families.

My colleague, Dr Brij Lal, was a great field scientist. Originally from Jaunpur Block of Tehri Garhwal, Uttarakhand, he is a botanist specialising in ethnobotany. Now settled in Dehradun, he completed his master's and PhD from MMPG College, Modinagar, affiliated with Meerut University. For his doctoral research, he worked under the guidance of a mentor from the CSIR–NBRI, Lucknow. We would go to plants in the fields and, as if in a meditative state, discuss how plants survive harsh environments and draw parallels to mechanisms applicable to other organisms, including humans. During these lonely and long sojourns, we developed theories about the pathways through which they perform varied functions, their capacity to adapt to changing environments, and their transfer from one generation to another. Many of these were never published but remained in our minds as vague thoughts. However, it later turned out to be true, as advances in agriculture, biotechnology, and the Ayurvedic system of medicine validated our early reflections on adaptation and survival across species, while also contributing to enhanced disease prevention and treatment strategies across different species.

For example, photosynthesis in plants involves the conversion of light energy into chemical energy stored in glucose molecules, primarily through the light-dependent and light-independent (Calvin cycle) reactions. The key molecules involved include chlorophyll, adenosine triphosphate (ATP), a molecule that carries energy within cells, nicotinamide adenine dinucleotide phosphate (NADPH), a crucial coenzyme involved in various metabolic reactions, particularly in photosynthesis and cellular metabolism, and glucose. Cellular respiration in humans (and other animals) is the process by which glucose is broken down to produce ATP, the cell's primary energy currency. This process occurs in mitochondria and involves glycolysis, the citric acid cycle, and the electron transport chain.

Both processes —in animals and plants —are linked in the planet's biological energy flow: photosynthesis produces glucose and oxygen that humans and other animals consume. In turn, humans perform cellular respiration to generate energy from food, releasing carbon dioxide and water as byproducts. These processes are essentially inverse and demonstrate a fundamental biological symmetry across plant and human physiology.

I have no doubt that plants hold the key to the basic secrets behind all life forms in the easiest manner for whoever wants to seek them. This knowledge, as a template or a guide map, could lead to the development of new biotechnologies, such as crops engineered for better drought or cold tolerance, pharmaceuticals derived from plant compounds with protective properties, and innovations in protecting human health from ultraviolet (UV) damage or extreme temperatures.

Moreover, understanding these adaptations deepens our comprehension of biological resilience and evolution, potentially revealing fundamental principles applicable to survival in other harsh environments, including space exploration or climate change resilience. In essence, high altitude plant research may hold the key to unlocking hidden secrets of life's endurance and adaptability across the planet and beyond.

We used to stay in a guest house at Kaza. Although electricity connections existed, frequent and prolonged power cuts made it as good as having no electricity supply. So we carried our generator to keep our laboratory running. The biggest hardship was frozen water in the toilets. Sleeping in the dark becomes a habit, and when it was summer months in Palampur, Kaza would still have freezing temperatures at night. During my evening walk, I used to look at the starry sky at night, which appeared too low, like a roof, and I could read all the star constellations as paintings in the sky, starting with the never changing position of the North Star, which we call *Dhruva Tara*, and the Saptarishis constellation, also known as the Big Dipper or Ursa Major, moving around it in the counterclockwise direction as the night progresses as Earth spins from west to east.

Two students joined me, Narinder Kumar and Dhiraj Vyas, to pursue a PhD with me. Together we observed and articulated how high-altitude desert plants exhibit a remarkable array of physiological adaptations that enable them to thrive in some of the most challenging environments on Earth, characterised by low temperatures, which can fluctuate dramatically between day and night, intense UV radiation from the sun due to the thinner atmosphere, lower CO_2 levels available to perform photosynthesis, limited availability of water, and nutrient-poor soils.

Our work attracted the attention of Dr V.L. Chopra, a pioneering figure and a doyen of Indian biotechnology. His extensive contributions to the field have helped shape the landscape of biotechnology research and development in India. Dr Chopra's work has spanned various areas, including microbial biotechnology, plant biotechnology, and genetic engineering, laying the foundation for many scientific advancements in the country. He took all the trouble of visiting our research station and set up one experiment. Unfortunately, while returning, a landslide left him stranded, and he had to walk 60 km and sleep in the dark, unprotected places. He was 64 years old at that time and considered a national asset for his knowledge and experience.

Once, my colleague Dr Surinder Kumar Vats, a 37-year-old plant physiologist, fell sick at Kaza and his lungs got filled with water. High-altitude pulmonary oedema (HAPE) is a serious and potentially life-threatening illness. HAPE occurs when fluid leaks from the blood vessels into the air sacs of the lungs due to rapid ascent to high elevations, especially above 8,000 feet. This excessive fluid accumulation hinders the lungs' ability to oxygenate the blood properly, resulting in severe breathing difficulties. He had to abandon his trip and return due to a medical emergency. Dr Vats, who later retired after a respectable career, was then a bright young scientist. He remained devoted to plant physiology, ever helpful to colleagues, and respected for his sincerity and gentle nature. Even after retirement, he continued his commitment and was re-employed as a CSIR Emeritus Scientist. No wonder, the families of CSIR-IHBT people working in Lahaul-Spiti used to consider their safe return as a reincarnation and perform *Puja*.

To cope with water scarcity, these plants often develop extensive or deeply penetrating root systems that can access underground moisture reserves. Their tissues are also adapted to reduce water loss; for instance, many possess thick, leathery, or hairy leaves that minimise transpiration by creating a physical barrier or trapping moisture. Moreover, some species have evolved waxy cuticles that serve as a protective barrier against water evaporation and environmental stress.

In response to the wide temperature fluctuations and freezing conditions, high altitude desert plants produce cellular antifreeze proteins and heat-shock proteins that stabilise cell membranes and prevent damage caused by temperature extremes. These biochemical strategies ensure cellular functions continue efficiently despite harsh conditions. UV protection is another critical adaptation, as the high elevation results in increased exposure to harmful radiation. To mitigate UV damage, plants synthesise protective pigments such as anthocyanins and flavonoids, which absorb and dissipate UV energy. Additionally, a thick cuticle layer and pigmented tissues serve to shield underlying cells from UV harm.

We now start publishing our research in international journals such as Journal of Plant Physiology (Vols. 158 and 164),

Photosynthesis Research (Vol. 88), and later our work also appeared in Plant Physiology (Vol. 177), Plant Biotechnology Journal and several others. Photosynthetic efficiency is crucial for survival in such environments, and many of these plants, such as cacti [family *Cactaceae* (various genera, e.g., *Opuntia* Mill.)], pineapple (*Ananas comosus* (L.) Merr.), Agave, etc., utilise Crassulacean Acid Metabolism (CAM) photosynthesis. This specialised form of photosynthesis allows the plants to open their stomata at night, when temperatures are cooler and humidity is higher, thus significantly reducing water loss. During the day, they close their stomata and carry out most of their photosynthesis using stored organic acids, which minimises dehydration. About 85% of plants, such as wheat (*Triticum aestivum* L.), rice (*Oryza sativa* L.), barley (*Hordeum vulgare* L.), soybean (*Glycine max* (L.) Merr.), and cowpea (*Vigna unguiculata* (L.) Walp.), fix carbon dioxide through photosynthesis, which is popularly known as the C3 carbon fixation pathway, in which the first stable compound formed is a three-carbon molecule called 3-phosphoglycerate. There is yet another group of plants called C4 plants, such as sugarcane (*Saccharum officinarum* L.), maize (*Zea mays* L.) and sorghum (*Sorghum bicolor* (L.) Moench), wherein the first stable compound formed is a four-carbon molecule called oxaloacetic acid. Generally, C4 plants thrive in hot and dry environments. Notably, C3 plants release up to 20-40% of the carbon fixed during photosynthesis as carbon dioxide, and also release ammonia through photorespiration. In contrast, such emissions are negligible in C4 plants, which gives them a yield advantage.

We brought out, for the first time, if I may say so, that C3 plants at high altitude refixed photorespiratory carbon dioxide and ammonia and hence emission loss of carbon dioxide and ammonia is negligible. While there are international efforts to convert C3 crop plants into C4 type, we offered a new approach which avoids this conversion, while offering a yield advantage. Well-established in their fields by now, Dr Narinder went to the Institute of Plant and Microbial Biology (Academia Sinica), Taipei, Taiwan, and Dr

Vyas joined CSIR-Indian Institute of Integrative Medicine (CSIR-IIIM) Jammu.

Late in my career, Dr Dharam Singh joined me as a postdoctoral fellow to scale up the production of SOD. Together, we helped the industry enhance enzyme production while giving Dr Dharam Singh valuable hands-on exposure to industrial enzyme research and innovation. Yet another postdoctoral fellow, Dr Rahul Jain joined me to work on plant-microbe interactions. Working at high altitude, we realised that plant adaptation in the harsh environment of high altitude is facilitated through symbiotic relationships with soil microbes, such as yeast, which modulates hormone metabolism. In hindsight, without even being aware of the larger plan, we were always preparing for more challenging assignments.

In 2003, I was selected for a DBT fellowship to work at Rothamsted Research, Harpenden, UK, with Prof. Christine H. Foyer for six months. I reached there in the first half of February, but already accustomed to living at a high altitude, the snow and winter there looked mild to me. Prof. Foyer is a renowned plant scientist known for her groundbreaking research in plant physiology, biochemistry, and stress biology. Her work primarily focused on understanding how plants respond to environmental stresses such as drought, salinity, and extreme temperatures, with particular emphasis on oxidative stress and reactive oxygen species. She has made significant contributions to elucidating the molecular and biochemical mechanisms that enable plants to adapt and survive in challenging conditions, including high-altitude and arid environments. Her research has important implications for improving crop resilience and food security in the face of climate change.

I worked at Rothamsted on molecular aspects of ascorbate metabolism in plants, and among the laboratory team was Guy Kiddle, a British national. Prof. Foyer used to work across nations, including South Africa, Argentina and so on. Ascorbate (vitamin C) plays a vital role in cellular processes related to cell death and disease resistance through its involvement in redox regulation, antioxidant activity, and enzyme function. Ascorbate acts as a potent antioxidant, neutralizing reactive oxygen species (ROS) that can induce oxidative stress and trigger apoptosis or necrosis. It modulates redox-sensitive signalling pathways, such as those involving NF-κB and p53, which govern cell survival and death and can influence epigenetic regulation by serving as a cofactor for dioxygenases involved in DNA and histone demethylation, affecting gene expression related to apoptosis.

For the first time, I felt like a molecular innovator with the yearning of a drug developer, as if a new mental birth. High ascorbate levels bolster the cellular antioxidant capacity, reducing oxidative damage from pathogens or stress. Ascorbate participates in the biosynthesis of defence-related compounds, like lignin and phytoalexins in plants, enhancing disease resistance. It modulates immune responses by influencing the function of enzymes and signalling molecules involved in pathogen recognition and defence pathways.

Ascorbate helps neutralise oxidative stress caused by free radicals, protecting cells and tissues from damage associated with ageing, inflammation, and chronic diseases such as cancer and cardiovascular illnesses. It boosts immune function by supporting the production and activity of white blood cells, enhancing the body's ability to fight infections. Adequate ascorbate levels promote faster wound healing and tissue repair. Can high-dose vitamin C selectively induce apoptosis in certain cancer cells and reduce tumour growth? I wondered.

Back home, Dr Ahuja secured funding to create a proper research station at Lahaul, called the Centre for High Altitude Biology (CeHAB), with facilities at Tandi (3,100 m, for office and laboratory) and Ribling (3,450 m, largely for field experiments). The Himachal Pradesh government allotted the land. It was a well-accessible area on the way to Leh. We recruited Dr Amit Chawla (a PhD from Guru Nanak Dev University, Amritsar) and Dr Rakshak Kumar (a PhD from North Eastern Hill University, Shillong) to work at the station. Fortunately, two locals, Dr Kiran Devi (a PhD from Guru Nanak Dev University, Amritsar) and Dr Ashok Kumar (a PhD from Kumaon University, Nainital, Uttarakhand), also joined us, and we now have an active Bioresource Laboratory of CSIR-IHBT– the first in India for sure, and no known counterpart in the public domain elsewhere in the world. But our independent identity would take many more years to emerge.

Now half plant-physiologist and half-drug discoverer, I looked at the *Ratanjot* plantation, a hallmark of the Spiti region. Botanically known as *Arnebia euchroma* (Royle ex Benth.) I.M. Johnst., *Ratanjot* is a perennial herb in the Boraginaceae family, having a pink to purple flower. The plant has a root of blackish appearance externally, but blue-red inside, with a whitish core. The root produces a fine red colouring material, which has been used as a dye since antiquity. Interestingly, local people know that, in alkaline environments, alkanet dye is blue; in the presence of acids, it gradually changes to crimson. The colour is red in an acidic medium (pH 6.1), purple at a mildly alkaline medium (pH 8.8), and blue at a stronger alkaline medium (pH 10.0).

But what is even more interesting is that colours only exist in the human mind. Colours are not physical properties inherent in objects themselves but are experiences created by our minds. They exist as perceptions, interpretations, and sensations processed by the human brain based on the way our eyes and nervous system interpret light wavelengths. In essence, colours are mental phenomena—constructs of consciousness—that help us make sense of the visual world. Without human perception, colours would have no meaning or existence outside of our minds.

Chemically, dyes are complex organic compounds designed to bond with fibres, whether natural or synthetic. The bonding process can involve various types of interactions, including covalent bonds, hydrogen bonds, or van der Waals forces. These interactions enable the dye molecules to adhere firmly to the material, ensuring that the colour remains durable through washing and wear. Different types of dyes—natural dyes derived from plants or minerals and synthetic dyes manufactured through chemical processes—have distinct properties affecting their colour fastness, brightness, and application methods.

While the science explains the physical and chemical mechanisms behind the production of colour, it is important to understand that colours themselves do not exist as inherent properties within objects. Instead, colours are created by the way our eyes and brains interpret the light reflected from these dyed materials. In essence, colour is a perceptual phenomenon occurring within the human mind. The physical processes of dyeing produce the colour we see, but the experience and interpretation of that colour depend entirely on human perception. This distinction underscores that, scientifically, dyes work through chemical and physical principles, while colour remains a subjective and mental experience.

The dyestuff in the roots of the *Ratanjot* plant is soluble in alcohol, ether, and oils, but is insoluble in water. It is used to give colour to wines and alcoholic tinctures, to vegetable oils, and to varnishes. *Ratanjot* is also valued for its anti-inflammatory and cooling properties, making it beneficial for treating burns, wounds, and skin infections. It can help reduce inflammation, soothe irritation, and promote faster healing.

It is incredible how different parts of a single plant can produce hundreds of distinct molecules due to the complex biochemical processes that occur within the plant's various tissues and structures. Each part of the plant—roots, stems, leaves, flowers, seeds, and bark—has its own unique environment and function, which influences the types of chemical compounds it synthesises.

For example, leaves are often rich in chlorophyll, which is essential for photosynthesis, while flowers contain flavonoids and pigments that attract pollinators. Roots may produce alkaloids and other secondary metabolites that serve as defence mechanisms against pests and diseases. Bark and stems can contain lignin and resins, providing structural strength and protection. Seeds often contain stored nutrients and protective compounds that support the developing embryo.

These diverse molecules include alkaloids, terpenes, flavonoids, phenolics, tannins, and many others—each with specific functions that help the plant survive, reproduce, and adapt to its environment. The biosynthesis of these compounds involves complex enzymatic pathways, and different environmental conditions—such as light, temperature, and soil nutrients—can influence which molecules are produced and in what quantities.

Take, for example, Himalayan rhubarb, botanically called the *Rheum australe* D. Don. It is a stout herb plant with large, heart-shaped or orbicular leaves, and reddish-purple flowers. Its leaves can be as wide as 60 cm. Traditionally used as a purgative, astringent, tonic, and for treating ulcers. From the scientific perspective, it contains derivatives of anthraquinone molecule that has the formula $C_{14}H_8O_2$. Natural products such as emodin, aloe-emodin, rhein, chrysophanol, and physcion are all anthraquinone derivatives. It also has stilbenes (like piceatannol and resveratrol). I have no doubt that anthraquinones interfere with DNA replication and repair processes, making them effective in targeting rapidly dividing cancer cells.

One well-known example is doxorubicin (also known as Adriamycin), an anthraquinone derivative of microbial origin, widely used in chemotherapy for treating cancers such as leukaemia, lymphoma, and breast cancer. Doxorubicin works by intercalating into DNA strands, disrupting the function of the topoisomerase II enzyme, and generating free radicals that damage cellular components. Research continues into synthesising new anthraquinone derivatives and understanding their mechanisms of action, with the goal of developing more effective and less toxic anti-cancer therapies.

Relying solely on the synthetic production of these compounds can be costly and complex. Preserving the natural plants provides an authentic and sustainable source for extraction and research, supporting ongoing scientific and pharmaceutical advancements. Therefore, protecting and conserving anthraquinone-producing plants is not only about safeguarding biodiversity but also about securing the potential health benefits they harbour for generations to come.

When I used to stand at Rohtang Pass in the early 1990s, I could see lush green patches of this plant around, but now they have all disappeared. As roads are being laid and celebrated as development, such plants are becoming extinct. Even *Ratanjot* is disappearing. If these plants disappear, we risk losing a vital reservoir of chemical diversity that could contain compounds crucial for future medical breakthroughs. The loss of these plants would also result in a reduction in biodiversity, which is crucial for maintaining ecological balance and resilience.

The Himalayan mayapple, botanically called *Sinopodophyllum hexandrum* (Royle) T.S. Ying, is a big red flag. A perennial herb, known for its rhizomes used in anticancer drugs, is now considered an endangered species. The overharvesting of Himalayan mayapple for medicinal purposes, coupled with habitat loss due to deforestation, agricultural expansion, and illegal collection, has led to a decline in its natural populations. As a result, it has been classified as endangered by conservation authorities in several countries.

Conservation efforts are crucial to protect it, including sustainable harvesting practices, cultivation programs, and legal protections to prevent illegal collection. Preserving this species is vital not only for maintaining biodiversity but also for ensuring a continued supply of its valuable medicinal compounds. But the point is, who will do it is an issue. Scientific laboratories can only do their job of utilising them, but to save them from development, which is often blind to such nuances, is beyond them.

The newly constructed Rohtang Tunnel, popularly known as the Atal Tunnel, stands as a remarkable achievement in engineering and a vital strategic infrastructure project. Spanning approximately 9 km, it is the world's longest subterranean vehicular tunnel above 10,000 feet, providing all-weather connectivity to the remote and picturesque Lahaul-Spiti Valley in the Himalayas. This tunnel not only symbolises India's technological advancement and engineering expertise but also plays a crucial role in boosting regional development by ensuring year-round access, especially during harsh winter conditions when snow and landslides often block traditional routes.

However, while the Atal Tunnel is a testament to human ingenuity, it also raises serious concerns about the environmental impact on this fragile Himalayan ecosystem. The hills surrounding the region are inherently delicate, composed mostly of loose stones held together by minimal soil and mud. Historically, the area had minimal disturbance from heavy construction or large crowds, and the climate remained relatively stable with limited rainfall.

The rapid influx of tourists and builders due to improved accessibility and infrastructure development threatens to alter this delicate balance. Increased human activity, coupled with climate change, could lead to unpredictable weather patterns, including heavy rains and cloudbursts. Such changes would have devastating consequences, as the existing fragile terrain could easily become unstable, leading to landslides, soil erosion, and other natural disasters. These ecological disturbances could trigger catastrophic events, such as massive landslides or flooding, which would not only endanger lives but also cause irreversible damage to the region's biodiversity and natural resources.

Lahaul-Spiti is known for its unique alpine flora, wildlife, and traditional cultures, which have been preserved over centuries. Increased construction, vehicular traffic, and tourism lead to habitat destruction, deforestation, and pollution, threatening local plant and animal species. The delicate balance of high-altitude ecosystems is highly vulnerable to environmental disturbances, and the loss of biodiversity could have long-term ecological consequences.

Moreover, unregulated tourism and urbanisation can lead to soil erosion, water scarcity, and degradation of natural resources, further endangering the region's ecological integrity. It is crucial to promote sustainable development practices that balance economic growth with environmental conservation. Protecting the biodiversity of Lahaul-Spiti is essential to preserving its natural beauty, ecological stability, and cultural heritage for future generations, even as modernisation continues.

Our work was well received by the best in the world. In 2006, I was selected to work at the Ecological Genomics Institute (EcoGen) at Kansas State University in Manhattan, Kansas, United States. It was funded by Targeted Excellence, Kansas State University, United States. It had a two-year tenure with the family to work on ecological genomics, an interdisciplinary field that combines genomics, ecology, and evolutionary biology to understand how genomes influence ecological adaptations and interactions.

Ecological genomics research focuses on how genetic variation underpins adaptive traits in natural populations, enabling species to survive and thrive under changing environmental conditions. This is what my team had been doing in Lahaul-Spiti. By studying the genomic basis of ecological traits, we can now uncover the mechanisms of adaptation, speciation, and resilience in diverse ecosystems. This research would provide valuable insights into biodiversity conservation, ecological dynamics, and the impacts of climate change on natural populations.

At EcoGen, my journey was enriched by working with Dr Loretta C. Johnson and Dr Jyoti Shah. Dr Johnson, a plant ecologist with a deep understanding of prairie ecosystems, had an infectious passion for how grasses adapt to changing climates. My work included studies on big bluestem (*Andropogon gerardii* Vitman), a dominant tallgrass prairie species. Dr Shah, with his expertise in plant NextGen Biology and defense mechanisms, provided another dimension. Both were not just colleagues, they were mentors who welcomed me into their world with warmth and generosity.

In August, I once again returned to India on a one-month leave to take Richa and the children to Manhattan, confident that staying together in the United States would open new vistas for all of us. However, fate had other intentions, ready to test the old adage: "Man proposes, God disposes." Despite our intentions and preparations, an unforeseen event was looming, ready to alter the course of our lives. On the morning of August 22, my day took a sudden and alarming turn. I experienced unexplainable uneasiness as if my left hand was congested with blood. The symptoms crept in unexpectedly, casting a shadow of concern over my morning. What followed was a series of moments filled with worry and uncertainty, reminding me that life's true plans are often beyond our control.

Richa called Dr H.P. Singh, my colleague living next door, and he took me to the local hospital, which had no facility beyond a basic ECG machine, that too kept on the second floor and I had to climb two flights of stairs in great pain. When the ECG confirmed a heart attack, I was taken to Fortis Hospital, Chandigarh, by road, where an angioplasty was carried out the next day. There was no way I could get admission in the famed PGI-Chandigarh hospital and receive free treatment, which I was entitled to, and the hospital bill exceeded ₹3 lakhs, which I paid out of pocket, with only partial reimbursement as per the extant rules. My parents and siblings came to see me, and there was no one in the little community of CSIR-IHBT who had not shared my pain and anguish.

There was no way I could join the EcoGen again. Though Dr Johnson tried a few times, I made up my mind to let it go. Richa said with her characteristic half-smile that maybe fate had other plans than going to the USA. One of the highlights of my stay at EcoGen was my exposure to the Konza Prairie Long-Term Ecological Research (LTER) site, one of the world's most intensively studied tallgrass prairie ecosystems. Here, studies are conducted across multiple ecological levels (organismic, population, community, and ecosystem), as well as across spatial scales (from plot-level to watersheds and regional landscapes) and temporal scales (ranging from days to decades).

Spreading across the rolling hills of Kansas, LTER plots were living laboratories where decades of data had been collected on fire regimes, bison grazing, rainfall manipulations, and nutrient dynamics. Walking those plots, I realized how essential such long-term experiments are for understanding adaptation and resilience.

While at Palampur, this experience gave us both the vision and the confidence to establish long-term ecological research plots in Himachal Pradesh. In the Himalayas, our own plots would one day help unravel the complex interplay between genes, environment, and adaptation across years and decades, knowledge that is vital for climate change studies.

One evening, Dr Ahuja came to our house. He said philosophically, "*Kakke*, you were searching for the secret of life in plants, and the unseen hand of destiny showed you a higher order of reality where human beings are merely a cog in one wheel of a giant machine. But remember my words, whatever happens, there is always a bigger reason. Wait and see in faith."

•••

My parents, Shri Mahendra Krishnan and Smt. Shakuntala Devi. Parents' love is the first shelter, the steady light, and the lasting compass of life.

I grew up in the Bhilai Steel Plant township—a place of order and harmony, where life moved in an organised and peaceful rhythm.

With my father and grandfather, Shri Ved Raj. Each of us is a vessel of continuity—genetic material intertwined with the experiences of those who came before, alive in our consciousness.

I graduated in Science from Lucknow Christian College, Lucknow—a place that nurtured my curiosity and shaped the path I would walk.

*At College of Basic Sciences and Humanities, Pantnagar,
I first tasted the quest for science—immersing myself in the
mysteries of life through an MSc in Plant Physiology.*

*I was fortunate to find my place at the
ICAR-Indian Agricultural Research Institute (ICAR-IARI) for a PhD,
the epicentre of agricultural sciences in this part of the world.*

Prof. Suresh Kumar Sinha, my PhD guide, was not only a mentor in science but the guiding soul who set my journey on its true course.

I joined CSIR Complex Palampur, and dedicated the entirety of my professional career to its growth and mission.

The Complex grew to become the CSIR-Institute of Himalayan Bioresource Technology (CSIR-IHBT). From a humble beginning to a world-class institute, I mirrored it in my life.

At Texas Tech University, Lubbock, I calibrated myself to global standards and broadened my horizons.

With Dr Abhraham Blum (left) and
Prof. Henry T. Nguyen (centre) at Texas Tech University.

I could earn my place in the team of the best of the best scientists during my tenure at Rothamsted Research, Harpenden, England, which is one of the oldest agricultural research institutions in the world.

My biochemistry teacher, Prof. Govind Krishna Garg at Govind Ballabh Pant University of Agriculture and Technology, became a dear friend and collaborator.

With Dr Paramvir Singh Ahuja in Cologne, Germany. In the background, one can see the Hohenzollern Bridge's "love locks" on the Rhine River. Traditionally, couples engrave a padlock with their names, attach it to the bridge's railings, and throw the key into the River Rhine to symbolise their lasting love.

Crossing Rohtang Pass (13,058 feet) on the eastern end of the Pir Panjal Range into the cold desert of Lahaul and Spiti Valley marked my own personal transformation from a learner to a doer.

It was time to apply the global learnings in my own country. The creation of the Centre for High Altitude Biology (CeHAB) of CSIR-IHBT in Tandi, Lahaul & Spiti was the first bold step.

4

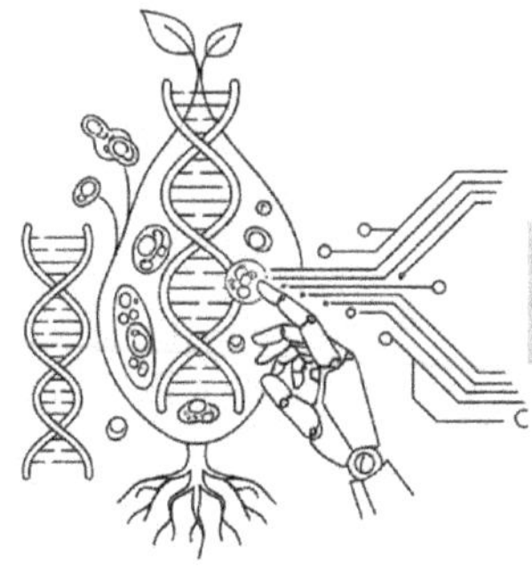

INTO THE NANO

Within the smallest particles lies the vastest promise; the invisible holds the power to transform the visible world.

I had avoided visiting Lahaul-Spiti for about a year, during which I refined my theory on carbon dioxide (CO_2) and ammonia (NH_3) refixation. Now, it was time to test what advantages the theory would offer. Carbon fixation refers to the process by which local plants absorb CO_2 from the atmosphere and convert it into organic compounds during photosynthesis. This process occurs during photosynthesis, primarily through a process known as the Calvin cycle. Inside the chloroplasts, the enzyme ribulose 1,5-bisphosphate carboxylase (Rubisco) captures CO_2 and combines it with a five-carbon sugar called ribulose 1,5-bisphosphate (RuBP). This creates an unstable six-carbon compound that quickly splits into two three-carbon molecules, 3-phosphoglycerate (3-PGA). Energy from ATP and NADPH (produced during the light-dependent reactions of photosynthesis) is utilised to convert 3-PGA into glyceraldehyde-3-phosphate (G3P), a three-carbon sugar. Some G3P molecules leave the cycle to form sugars, while others regenerate RuBP, allowing the cycle to continue.

Nitrogen is a core element of chlorophyll, the pigment responsible for photosynthesis. Retaining nitrogen ensures plants can produce enough chlorophyll, supporting efficient photosynthesis. Nitrogen is a component of nucleotides, which are essential for the synthesis of DNA and RNA. This supports cell division and growth. Retaining nitrogen helps plants develop healthy leaves, stems, roots, and reproductive structures, promoting better growth and higher yields. Adequate nitrogen levels improve plant resilience against environmental stresses, pests, and diseases.

We consider this an adaptation in response to low partial pressure of gases [what matters to plants or humans is the partial pressure of CO_2 (for photosynthesis), and O_2 (for respiration), respectively; air is a mixture of various gases, and the share of pressure exerted by each gas is called partial pressure] at high altitude, which occurs due to a reduction in atmospheric pressure with an increase in altitude, for example, about a 50% reduction at 5,500 m altitude. Humans increase the haemoglobin level to take care of the reduced partial pressure of O_2, and we found that plants deploy the two enzymes simultaneously. The advantage was that while the fixed product of Rubisco (3-PGA) goes towards the synthesis of glucose and other molecules, the fixed product of phosphoenolpyruvate carboxylase (PEPCase) serves as a carbon substrate to refix ammonia, using the CO_2 liberated by the photorespiratory process. When we introduced the refixation mechanism of carbon (CO_2) and nitrogen (ammonia form) in other C3 plants, *Arabidopsis thaliana* and *Brassica juncea* L. and published the results, it attracted international appreciation.

Winter dormancy in plants is a survival mechanism that allows them to withstand the harsh conditions of winter. As the temperature drops and the days become shorter, plants receive signals to enter a dormant state. During this period, their growth and metabolic processes slow down significantly to conserve energy and resources. Deciduous trees, for example, shed their leaves to reduce water loss and prevent damage from the cold and frost.

Physiological changes accompany dormancy; for instance, the production of protective substances like sugars and antifreeze proteins helps prevent cell damage caused by freezing temperatures. Additionally, chlorophyll levels decrease, making the foliage less active and further reducing metabolic demands. Many plants develop protective structures, such as thick bark or buds, that shield their vital tissues from the cold. Hormones like abscisic acid increase during dormancy, signalling the plant to halt growth and prepare for the winter. This state of suspended activity persists until environmental conditions become favourable again. When temperatures rise and daylight increases in spring, dormancy is broken, and growth resumes. Plants then begin to flower and carry on their life cycle.

In tea plants, winter dormancy is a crucial part of their annual growth cycle, especially in regions with cold winters. During the winter months, when temperatures drop and daylight hours shorten, tea plants enter a state of dormancy. In this period, the growth of new shoots and leaves slows down significantly or temporarily stops. This dormancy helps the tea plant conserve energy and resources, protecting it from damage caused by frost and cold stress.

Physiologically, the tea plant reduces metabolic activities such as photosynthesis and nutrient uptake. The formation of protective tissues and the accumulation of carbohydrates and antifreeze substances help the plant withstand freezing temperatures. Additionally, the plant's buds and stems develop a protective layer to shield the vital parts from cold damage. As spring approaches, with rising temperatures and longer days, the dormancy is broken, and the plant resumes active growth, producing new shoots and leaves that are harvested for tea production. This cycle of dormancy and growth is essential for maintaining healthy tea plants and ensuring high-quality tea leaves year after year.

During winter dormancy in tea plants, several specific metabolites are produced or accumulated to help the plant withstand cold temperatures and stress. One of the most important groups of these metabolites is sugars, including sucrose, glucose, and fructose. These sugars act as natural antifreeze agents, lowering the freezing point of the plant's cell sap and preventing the formation of damaging ice crystals within the cells. This accumulation of sugars helps to protect the integrity of cell membranes and other vital structures during freezing conditions.

In addition to sugars, tea plants produce antifreeze proteins such as dehydrins. These proteins play a crucial role in preventing ice crystal formation inside cells, offering further protection against cold damage. Proline, an amino acid, also accumulates during dormancy; it functions as an osmoprotectant, stabilizing proteins and cell membranes while helping to maintain osmotic balance within tissues. Moreover, polyphenols like catechins and other flavonoids increase in concentration, providing antioxidant protection against oxidative stress caused by cold temperatures.

Hormones like abscisic acid (ABA) are vital regulators of dormancy, exerting control over the synthesis of these protective metabolites. ABA promotes the induction and maintenance of dormancy, helping the plant conserve energy and restrict growth until conditions improve. Overall, these metabolites work synergistically to safeguard the tea plant during winter, ensuring its tissues remain viable and enabling the plant to resume growth when the environment becomes more favourable. Asosii Paul and Sewa Singh, who worked with me for their PhD, took a different angle on winter dormancy in tea. We asked why tea does not shed its leaves during winter, whereas others, such as poplar, do? We realised that tea avoids leaf fall through the genes related to fruit softening. Tea played a clever trick!

In tea, we also took the lead in deciphering the molecular control of catechins biosynthesis. Catechins are highly valued flavonoids responsible for giving antioxidant advantage to green tea and, after oxidation, are key determinants of the strength and colour of CTC (crush, tear, curl) tea. During this work, Kashmir Singh joined me for his PhD and carried out pioneering research on catechin biosynthesis in tea. We then extended the scope of metabolic engineering to other medicinal plants such as for iridoid biosynthesis in *Picrorhiza kurroa* Royle ex Benth. (a hepatoprotective plant with iridoids are responsible for this property), shikonin biosynthesis in *Arnebia euchroma* (*Ratanjot* roots give deep red coloured shikonin used in traditional medicines, as natural dye and in cooking), stevioside biosynthesis in stevia (sugar alternative; negligible calorie and safe for diabetic persons) and podophyllotoxin biosynthesis in Himalayan mayapple (it has precursor for semisynthetic molecule that is used for cancer treatment) among others. Such insights are essential to turn a living organism into a biofactory. On the strength of this body of work, I was invited, in July 2011, to attend the XVIII International Botanical Congress in Melbourne, Australia, where I delivered a talk in a session on "Plants in Human Health and Well-being."

In the timeless tradition of *Guru-Shishya Parampara*, learning is never one-sided; it is a shared journey where curiosity, insight, and dedication flow between teacher and student. I was privileged to walk this path with 31 bright minds, each bringing their own questions, creativity, and energy. Together, we explored the mysteries of plants, celebrated discoveries, and nurtured a collective passion for understanding life in all its richness. In this shared voyage, knowledge became a living exchange, and every insight illuminated the path for all of us.

In the realm of tea (*Camellia sinensis* (L.) O. Kuntze), some delved into drought responses and oxidative stress, others explored winter dormancy, while a group unravelled the complex pathways of catechin biosynthesis. Through patient exploration, transcriptome dynamics and polyphenol oxidase were revealed, deepening our molecular understanding of this vital crop. I was blessed as if to have brilliant students. Priti Sharma, Manjot Singh, Dr Kashmir Singh, Dhiraj Vyas, Lakhvir Lal, Arti Rani, Asosii Paul, Richard Chalo Muoki, and Sewa Singh were all committed to their work with the dedication and service of young monks in a monastery, each helping expansion of consciousness and leaving a distinct mark on our collective knowledge. Dr Kashmir Singh would later become Professor at Panjab University, Chandigarh and earned great renown, while the others too went on to carve their own distinguished paths.

Across medicinal and alpine plants, the inquiry expanded. Rishi Kumar Gara and Ravi Shankar Singh investigated *Arnebia euchroma*, while Priyanka Bhardwaj pioneered a new dimension in plant-microbe interactions, revealing how plants adapt through unseen collaborations. Tabasum Kawoosa, Parul Gahlan and Tanvi Sharma explored *Picrorhiza kurroa*, complemented by Harsharan Singh's molecular insights. The story of superoxide dismutase connected Rashmita Sahu's work on *Potentilla atrosanguinea* Lodd. ex Lehm. with Pradeep Kumar Bhardwaj, Arun Kumar and Shweta Guleria's studies on thermostable SODs from Himalayan flora and Sunny Patial's expression in *Pichia pastoris*. Sanjay Ghawana and Deep Mala illuminated the physiology of *Rheum australe*, while Anita Kumari studied *Sinopodophyllum hexandrum*.

The intricate dynamics of carbon and nitrogen metabolism at high altitudes were revealed through the collective efforts of Narinder Kumar, Anish Kaachra, and Mamta. Hitesh Kumar advanced understanding in *Stevia rebaudiana*, Monika Bhuria investigated universal stress proteins in *Arabidopsis*, Vishal Sharma studied transcription factors in apple, and Mohit Kumar Swarnkar examined prickle morphogenesis in rose. Through these varied inquiries, the richness of Himalayan flora and the resilience of plants under stress came vividly to life.

Together, these 31 journeys created more than research; they wove a living mosaic of curiosity, perseverance, and discovery. Each student's contribution illuminated new paths, and in this rhythm of shared exploration, the spirit of learning, growth, and wonder was fully realised. Through their work, the legacy of Himalayan plant biotechnology continues to grow, inspired by inquiry, grounded in nature, and sustained by the enduring bond between teacher and student.

In April 2009, I visited Forschungszentrum Jülich in Germany, one of the largest and most advanced national research institutions in Europe. The visit was a turning point in my understanding of how plant sciences were entering a new era of precision and scale. At Jülich, I had the privilege of meeting Prof. Dr Ulrich Schurr, then Director of the Institute of Chemistry and Dynamics of the Geosphere, ICG-3 Phytosphere; Dr Anika Wiese-Klinkenberg, Scientist at IBG-2: Plant Sciences; and Dr Heike Schneider, Senior Scientist and Coordinator of the Indo-German collaboration at IBG-2: Plant Sciences.

What struck me most during my interactions there was their collective commitment to integrating physics, biology, and engineering into the study of plants. In particular, they emphasised the emerging importance of high-throughput phenotyping as a tool to analyse large populations of plants in a relatively short span of time. Rather than relying solely on traditional field observations or small-scale experiments, phenotyping enables researchers to systematically measure plant growth, architecture, stress responses, and physiological functions under both controlled and field conditions. This shift promised to dramatically accelerate breeding programs, enabling scientists to link genetic potential with observable traits in real environments.

The facility itself was a marvel—automated imaging systems, climate chambers, and sensor-based platforms working in concert to capture data points invisible to the naked eye but critical to advancing plant research. I realised that phenotyping was not merely about efficiency; it represented a paradigm shift in how we understood plants—as dynamic systems interacting with their environment, not just static units of study. This visit impressed upon me how essential it would be for India and other developing countries to adopt such cutting-edge approaches, especially in the face of climate change and the urgent need to build climate-resilient crops.

Along with phenotyping, nanotechnology can aid in developing cold-resistant plant varieties by facilitating gene editing and delivery of dormancy-associated genes or regulators. Nanoscale carriers can help enhance the expression of genes responsible for cold tolerance, antifreeze protein production, and antioxidant accumulation, leading to more robust plants capable of withstanding harsh winter conditions. Overall, integrating nanotechnology into the study and management of winter dormancy opens up new avenues for improving plant survival, productivity, and adaptation in changing climates. I returned to Forschungszentrum in February 2013 and became familiar with the area. I also visited Berlin on this trip and saw its iconic landmarks like the Brandenburg Gate and the Reichstag Building, where German and European history was made.

In 2017, we had a review by the Parliamentary Standing Committee on Science & Technology related to the Environment and Forests departments. Later they also visited the CSIR-IHBT to verify our on-the-ground work. Among the various projects presented, our pioneering work on low-altitude apple cultivation particularly caught their imagination. The idea resonated strongly with the parliamentarian who hailed from Mizoram and was seeking viable alternatives to expand horticultural opportunities in their region.

We took them around our laboratories, showing them the virus-free rootstock developed through painstaking research and explaining how it could be transplanted in Mizoram and other hill states, thereby opening the door for apple orchards in landscapes where traditional varieties could not thrive. To bring the science from theory to tangible proof, we served them freshly harvested apples of the cultivars 'Anna' and 'Dorsett Golden.' These varieties, especially suited to areas with limited chilling hours, can successfully break dormancy and flower at or below 7 °C—conditions far more attainable in the northeastern hills than those required by conventional temperate apples.

The parliamentarians tasted the crisp, juicy fruits, their expressions shifting from curiosity to delight. In that moment, the old adage—"the proof of the pudding lies in the eating"—came alive before us. Science had spoken not through data or charts, but through the simple sweetness of an apple, carrying within it the promise of new livelihoods and expanded horizons for farmers in India's northeast.

Low altitude apple cultivation benefits from the generally warmer temperatures found at lower elevations. These temperatures influence key physiological processes such as flowering, fruit development, and ripening. While warmth can accelerate growth and lead to earlier harvests, excessively high temperatures may adversely affect fruit quality and yield. Therefore, selecting the appropriate variety that can thrive in warmer conditions is essential for successful cultivation in low-altitude regions.

Another important factor is the number of chilling hours, which are cold periods required to break dormancy and initiate flowering. Low altitude areas often experience fewer chilling hours compared to higher elevations. To address this, farmers might need to select cold-tolerant apple varieties or employ techniques to artificially increase chilling hours, ensuring healthy and synchronised flowering. Proper soil conditions also play a critical role; fertile soils with good drainage and optimal pH levels support vigorous tree growth and efficient nutrient uptake, leading to healthy fruit production.

Adequate water availability and humidity levels are vital for maintaining healthy apple trees, especially in low-altitude regions where conditions can sometimes favour excessive moisture. Proper irrigation, combined with good air circulation and drainage, helps prevent diseases related to excess moisture. Additionally, while the ample sunlight typical of low-altitude areas benefits photosynthesis and fruit ripening, it can also increase the risk of pest and disease outbreaks. Integrated pest management and disease control strategies are therefore essential to protect the crop and ensure a healthy yield.

Successful apple cultivation at low altitudes hinges on managing a variety of environmental and biological factors. Optimising temperature conditions, ensuring appropriate chilling hours, maintaining soil health, managing water supply, and controlling pests collectively contribute to the productivity and quality of apples grown in these regions. With proper practices, low-altitude orchards can be highly productive and yield high-quality fruit. Our variety would take three years to start bearing fruit after planting.

Proper care, including adequate watering, fertilisation, pruning, and pest management, can influence the time it takes for an apple tree to reach fruiting maturity. Flowers in the second year must be plucked and not allowed to become fruit.

For the first time, I felt that our work was recognised for its societal potential, and a political leader believed in it. We sent planting material of apple by trucks to Mizoram, a 2,900 km road journey from Palampur to Aizawl via Guwahati would take five days. Of course, people travelled by air. Later, we cultivated apples in the Ukhrul district of Manipur. In the Northeast, apples arrive from Myanmar as apples from Kashmir do not survive transport time and rot along the way. Our scientific work created livelihood for local people, who sold these apples at ₹200 a kg, at par with Myanmar apples and made a good profit in the process.

Returning to nanoscale studies, reducing agents play a crucial role in the synthesis of nanoparticles. Substances such as sodium borohydride ($NaBH_4$), hydrazine, and ascorbic acid are commonly used to reduce metal ions to their metallic form. These agents help control the size, shape, and uniformity of the nanoparticles during the chemical reduction process.

Stabilising or capping agents are also vital in nanoparticle synthesis to prevent agglomeration and ensure stability. Molecules like citrate, polyvinylpyrrolidone (PVP), and polyethene glycol (PEG) are often employed for this purpose. They adhere to the surface of nanoparticles, providing a protective layer that maintains dispersion and influences surface properties.

Solvents such as water, ethanol, and other organic liquids serve as media where nanoparticle synthesis occurs. The choice of solvent can affect the size, morphology, and composition of the nanoparticles. Surfactants like cetyltrimethylammonium bromide (CTAB) are also used to direct the growth and shape of nanoparticles by interacting with their surfaces during synthesis.

Chelating agents such as ethylenediaminetetraacetic acid (EDTA) help regulate the synthesis process by binding to metal ions, controlling their availability and preventing unwanted reactions. Additionally, pH adjusters like hydrochloric acid (HCl) or sodium hydroxide (NaOH) are used to fine-tune the acidity or alkalinity of the reaction environment, which significantly influences nanoparticle formation and stability. These reagents collectively enable precise control over nanoparticle synthesis, ensuring the desired properties for various advanced applications.

I developed an RNA isolation kit for isolating RNA from difficult tissues, as well as those susceptible to quick RNA degradation. This kit was also for self-reliance and a symbol of pride in having our own that works better than any international brand. The kit has a specialised reagent set to streamline the process of isolating RNA while removing proteins, DNA, and other contaminants.

Typically, RNA isolation involves lysing the cells or tissues with a lysis buffer that inactivates RNases and releases RNA. The sample is then processed through a series of steps, including the binding of RNA to a silica membrane or magnetic beads, washing to remove impurities, and elution to obtain purified RNA. Many kits also include enzymatic treatments or proprietary reagents to improve yield and purity. Our kit was very simple, which can be adopted by any NextGen Biology lab, even with minimal facilities. It does not involve any silica membrane or magnetic beads, but relies on regular reagents available in the lab.

RNA isolation kits are available from various manufacturers, tailored for different sample types and downstream applications, such as reverse transcription-polymerase chain reaction (RT-PCR), Northern blotting, and sequencing. The choice of kit depends on factors such as sample amount, purity requirements, and the intended use of the isolated RNA. Using a reliable RNA isolation kit ensures that high-quality, intact RNA is obtained efficiently for subsequent NextGen Biology experiments. International Centre for Genetic Engineering and Biotechnology (ICGEB) alone bought thousands of litres of RNA isolation solution from our institute. Dr Girish Sahni, former Director of the Institute of Microbial Technology in Chandigarh and later Director-General of CSIR, used to compare it to Mahatma Gandhi's Salt March in

Dandi in 1930, as it ushered India into NextGen Biology, breaking the prohibitive cost barrier.

I was appointed as the Director of CSIR-IHBT in June 2015. On the very first day, DG, CSIR Dr M.O. Garg called me over the phone and asked me to drive down to Dehradun. A renowned chemical engineer, Dr Garg had been the Director of the CSIR-Indian Institute of Petroleum (CSIR-IIP) in Dehradun for 13 years. I barely signed my joining papers at Palampur and proceeded to Dehradun. It took us nearly twelve hours to cover the 400 km distance, but I reported there the next morning. It was a two-day "*Chintan Shivir*" (brainstorming session) for CSIR labs, led by then-Union Minister for Science & Technology and Earth Sciences, Dr Harsh Vardhan. His deputy, Minister of State for Science & Technology and Earth Sciences, Y.S. Chowdary, was also there.

Each of the laboratory heads made a presentation about their achievements in transferring products and processes developed by them to industry. They also disclosed the patents awarded. The key outcome was the Dehradun Declaration, where all CSIR labs committed to becoming self-financing within two years. A thirty per cent budget must be earned. When Dr Harsh Vardhan said, "I have full confidence in our scientists. With proper encouragement and a focused approach, I am sure India will achieve the status of the world's greatest scientific powerhouse," I felt it was my responsibility. That moment became a spark for me. I knew our institute needed more than just a strategy to earn revenues—it needed a vision that could carry us into the future. Drawing strength from the Himalayas around us, I saw a clear path: to use science as a bridge between the region's immense natural wealth and the nation's aspirations for growth.

It was from this conviction that our vision emerged: "To be a global leader on technologies for boosting bioeconomy through sustainable utilisation of Himalayan bioresources." This was not a routine statement but a pledge, an affirmation that our work would be both globally relevant and locally rooted. It was about creating opportunities where science, innovation, and sustainability could walk hand in hand, turning the unique bioresources of the Himalayas into engines of prosperity for India and the world. Looking back, I realise that Dehradun was not the end of a meeting; it was the beginning of a journey, one that would define the spirit and direction of our institute for years to come.

Out of this very journey, even the logo of CSIR-IHBT was born, a visual identity that carried within it the same essence as our vision. It was more than a symbol; it was a story in itself, capturing the strength of the mountains, the richness of biodiversity, and the promise of science to transform lives. Whenever I look at it, I see not just a design, but a reminder of the moment when conviction crystallised into vision, and vision into identity.

Shortly thereafter, Dr Girish Sahni took over as the Director-General, CSIR. Well-known for his contribution to the production technology for India's first indigenous clot-buster drug, natural streptokinase and recombinant streptokinase, Dr Sahni was a scientist of a different mettle. A PhD in Biochemistry from the Indian Institute of Science, Bangalore and work experience at the University of California, Santa Barbara, USA, the Rockefeller University, New York, and Albert Einstein College of Medicine, New York, USA, he would jokingly call scientific paper writing an "infatuation with cellulose" and always strive for products. His high-efficiency process package for the production of natural streptokinase for the first time in India, called STPase, commercially launched by Cadila Pharma Ltd., Ahmedabad, in 2002, was a great step towards self-reliance in biotechnology.

Dr Sahni, Dr Sudeep Kumar, and I were in Mysuru on January 3, 2016, to attend the 103rd Indian Science Congress, which had a theme on "Science and Technology for Indigenous Development in India". Dr Sahni discussed indigenous plants and the Ayurvedic system of medicines and suggested evolving a program in a mission mode on the phytopharmaceuticals mission. Since CSIR-IIIM, Jammu had a legacy of working on plant-based pharmaceuticals, its Director, Dr Ram A. Vishwakarma, was assigned the responsibility to develop and lead the program.

Dr Sahni came to CSIR-IHBT in May 2016 and also called Dr Anil Tripathi, Director, CSIR-CIMAP, Lucknow. Dr Sahni stayed in Room No. 14 of our guesthouse and spent many hours envisioning a Mission around Aromatic Plants. Realising the legacy of CSIR-CIMAP in essential oil. He made Dr Tripathi responsible for the Aroma Mission, and a bigger responsibility was on my way.

To streamline and focus CSIR's research and development efforts, CSIR in 2018 adopted a "Theme" approach around eight key themes: (i) Aerospace, Electronics, Instrumentation & Strategic Sectors; (ii) Civil Infrastructure & Engineering; (iii) Ecology, Environment, Earth & Ocean Sciences and Water; (iv) Mining, Minerals, Metals and Materials; (v) Chemicals (including leather) and Petrochemicals; (vi) Energy (conventional and non-conventional) and Energy devices; (vii) Agri, Nutrition & Biotech; and (viii) Healthcare.

Each theme was to be led by a theme director, and I was elected the first theme director for the Agri, Nutrition & Biotech theme, which involved the participation of 37 institutes of CSIR. Once, Dr Tripathi and I were with Dr Sahni, who reminded us that CSIR Headquarters was always open to every laboratory—and to us as well. That day, however, he spoke with a special sense of trust. Aware that Dr Tripathi was Mission Director and I was Theme Director, he said much rested on our shoulders. Through our missions and themes, he believed we could not only make our work self-sustaining but also contribute nearly 30% to CSIR's overall earnings, supporting sister labs in turn. It was both his confidence in us and a reminder of our responsibility. His words still echo: "There should be no excuses—only progress."

In April 2016, I visited the National Institute of Medicinal Materials in Hanoi, Vietnam, as part of a CSIR delegation comprising Dr Anil Tripathi, Dr Srivari Chandrasekhar, Director of CSIR-Indian Institute of Chemical Technology, Dr Alok Kalra, Chief Scientist at CSIR-CIMAP, Dr Vibha Malhotra-Sawhney from CSIR Headquarters and Dr Anuradha Madhukar from the International

S&T Affairs Directorate of CSIR. It was a sort of star-assembly delegation covering all aspects of developing drugs from the plants.

Vietnam has a rich tradition of plant-based medicine utilising a wide variety of herbs, roots, flowers, and other natural ingredients for healing purposes. Popular plants include Gac (*Momordica cochinchinensis* (Lour.) Spreng.), known for its high antioxidant content and used to boost immunity and liver health; Dong Quai (*Angelica sinensis* (Oliv.) Diels), which is utilised for improving blood circulation and women's health; *Centella asiatica* (L.) Urban valued for its skin healing and mental clarity benefits; ginger (*Zingiber officinale* Roscoe), widely used to aid digestion and fight inflammation; turmeric (*Curcuma longa* L.), recognised for its anti-inflammatory and detoxifying properties; and bitter melon (*Momordica charantia* L.), used to help regulate blood sugar levels and purify the body. These herbal remedies are often combined with other traditional therapies like acupuncture and moxibustion, reflecting a holistic approach to health. Many of these herbal formulations have been passed down through generations, forming an integral part of Vietnamese healing practices and emphasising the natural, medicinal qualities of their indigenous plants.

With extensive discussions and information synthesis, the CSIR Phytopharmaceuticals Mission, launched in 2017, focuses on the cultivation, conservation, and development of phytopharmaceuticals, which are plant-based medicinal products. This mission aims to leverage India's rich biodiversity and traditional knowledge to advance the phytopharmaceutical industry. It involves various CSIR labs working on different aspects of phytopharmaceutical development, from identifying and cultivating medicinal plants to developing technologies for extracting and processing them into usable forms.

When I briefed him about it, Dr Ahuja, who was living alone after retirement, felt very happy. Besides being the CSIR-IHBT Director, Dr Ahuja also served as Director, CSIR-CIMAP, and has been a scientific mentor to CSIR-NBRI for some time. He affectionately held my hand and said, "Sanjay, consider cultivation of right plants and good manufacturing practices (GMP) grade extracting facility as the two feet of your mission. If both are not equally strong, you will not go far." He also gave his tips to promote the cultivation of high-demand medicinal plants and the conservation of endangered species, and on developing technologies for producing GMP grade medicinal plant extracts. Dr Ahuja passed away on January 22, 2017. His sister, Rajinder Kaur Bhalla, took care of everything, until his son, Dr Jaswinder Singh and daughter Kamal arrived in Mohali for the last rites. Dr Sahni also attended the cremation.

Determined and prepared to apply nanotechnology, I started demystifying and creating awareness amongst the youngsters. Nanotechnology is essentially a tool to manipulate what happens at the nanoscale of matter, that is, the size between 1 and 100 nanometres. One nanometre is a billionth of a meter. My stand was that nanotechnology is nothing new, but the scientific validation of what was traditionally practised based on evidence, but without understanding. No wonder that the use of plants as medicine is not new; in fact, it is older than recorded history. Even in the modern world, 80% of people use herbal medicine for primary health care. It was time to know the reason and live in light.

The plant kingdom has remained as an inexhaustible reservoir of traditional medicines and also a starting point for modern medicines. India is known for its traditional medicinal systems—Ayurveda, Siddha, and Unani. These systems are mentioned in the ancient Vedas and other scriptures. Herbal medicine has remained an important alternative to modern allopathic medicine globally. Since ancient times, a number of Indian medicinal plants have been used globally. The problem is with safety. For most herbal drugs, the efficacy has not been proven and the quality is not assured. And it is here that nanotechnology is making the most difference.

Taking a cue from the global trends and opportunities in plant-based medicines, the Government of India has amended the Drugs and Cosmetics Act, 1940 and Rules 1945 to include phytopharmaceutical drugs as a new category and made regulatory provisions for phytopharmaceuticals, which is now defined as a purified and standardised fraction with a defined minimum of four bio-active or phytochemical compounds (qualitatively and quantitatively assessed) of an extract of a medicinal plant or its part, for internal or external use of human beings or animals for diagnosis, treatment, mitigation, or prevention of any disease or disorder but does not include administration by parenteral route. This has given a new hope for innovations and development of new drugs from botanicals in a scientific way, and would help in the global acceptance of the use of herbal products by the modern medical profession.

Concerned about India's approach to organising its herbal products, several international pharmacopoeias—including the British, Japanese, US, German Commission E, and Chinese Pharmacopoeia—have been established, all emphasising the importance of maintaining quality standards within their respective countries. Therefore, it is crucial to subject Indian medicinal herbs to rigorous modern scientific testing and establish standards to ensure quality and enhance global competitiveness.

Standardisation and quality control of raw materials and herbal preparations are essential for internationally accepted botanical products. When the active principles are unknown, marker substances are used for analytical purposes; however, these markers often lack validation for their role in the therapeutic effects of the herbs. Factors such as extraction methods and contamination with microorganisms, heavy metals, pesticides, and other pollutants can also compromise the quality, safety, and efficacy of herbal drugs. Consequently, pharmaceutical companies prefer cultivating plants over wild-harvested ones, as cultivated plants exhibit fewer variations in phytoconstituents. Moreover, cultivation allows for monitoring of key secondary metabolites, aiding in identifying the optimal harvesting period.

The primary laboratory for this initiative was CSIR-Indian Institute of Integrative Medicine (CSIR-IIIM), Jammu. Collaborating laboratories included CSIR-Central Institute of Medicinal and Aromatic Plants (CSIR-CIMAP), Lucknow; CSIR-

Central Drug Research Institute (CSIR-CDRI), Lucknow; CSIR-Institute of Himalayan Bioresource Technology (CSIR-IHBT), Palampur; CSIR-Indian Institute of Chemical Biology (CSIR-IICB), Kolkata; CSIR-National Botanical Research Institute (CSIR-NBRI), Lucknow; CSIR-Northeast Institute of Science and Technology (CSIR-NEIST), Jorhat; and the CSIR-Unit for Research and Development of Information Products (CSIR-URDIP), Pune.

The CSIR Phytopharmaceutical Mission aims to revolutionise the medicinal plant sector by implementing targeted interventions in agriculture, processing, and drug discovery. Its goal is to foster growth in the phytopharmaceutical industry and promote rural employment by improving the availability of high-demand medicinal plants through captive cultivation. This aligns with the broader objective of supporting the development of traditional Indian medicines for both domestic and international markets.

The Mission aims to enhance the availability of high-demand medicinal plants, primarily through cultivation, to meet the needs of both the global and domestic industries involved in traditional Indian medicine preparations. It seeks to prevent the depletion of these plants in their native habitats by identifying elite germplasm and conserving it through cultivation and gene banks. The mission will develop improved varieties and agro-technologies to boost productivity and profitability per unit area, including utilisation of regions affected by abiotic stresses such as drought, salinity, flooding, and shade. Additionally, it will focus on developing chemical processes for producing standardised extracts and enriched fractions from selected medicinal plants. These technologies will be transferred to entrepreneurs to promote value addition and encourage the export of processed materials instead of raw plant parts. A dedicated effort will also be made to translate promising clinical leads identified in various CSIR laboratories into phytopharmaceutical drugs, following regulatory pathways through AYUSH and DCG(I), for both domestic and international markets.

This part is done already and done well. The time is now to apply nanotechnology, which offers innovative solutions to manipulate matter at a very small scale, typically below 100 nanometres. When applied to phytopharmaceuticals, nanotechnology can enhance the delivery, absorption, and overall efficacy of plant-derived compounds. Nanocarriers such as nanoparticles, liposomes, and nanoemulsions can be used to encapsulate phytochemicals, protecting them from degradation and improving their solubility in bodily fluids. This targeted approach allows for more precise delivery to specific tissues or cells, potentially increasing therapeutic outcomes while reducing side effects.

The integration of nanotechnology with phytopharmaceuticals provides several key benefits. It can increase the solubility of poorly water-soluble plant compounds, enhance their absorption and bioavailability within the body, and allow for controlled or sustained release of active ingredients. Additionally, targeted delivery systems can minimise toxicity and side effects by concentrating the therapeutic agents where they are needed most. Overall, this multidisciplinary approach holds promise for the development of advanced herbal medicines that offer improved treatment efficacy and patient outcomes.

The visit to the Institute of Traditional Medicine, National Yang-Ming University, Taiwan, in July 2019 came at a great time. We have articulated the road ahead on phytopharmaceuticals to be a world leader, and this was a perfect platform to announce our intent. While Indian herbal medicine, primarily Ayurveda, emphasises balancing the body's three doshas—*Vata*, *Pitta*, and *Kapha*—using a variety of herbs, spices, and natural remedies, Chinese herbal medicine, rooted in Traditional Chinese Medicine (TCM), focuses on restoring balance within the body's vital energy, or Qi, through a combination of herbs, acupuncture, and dietary therapy. Their key herbs, like ginseng (*Panax ginseng* C.A. Meyer), dong quai (*Angelica sinensis* (Oliv.) Diels), liquorice (*Glycyrrhiza glabra* L.), and ginger (*Zingiber officinale* Roscoe), are commonly used to support energy, blood circulation, and immune function. Complex herbal formulas tailored to specific patterns of disharmony are common.

The future of medicine for currently incurable diseases is increasingly centred around plant-derived molecules, particularly those discovered in unique environments such as high altitudes. These extreme conditions often lead to the development of plants with potent bioactive compounds that have evolved as survival mechanisms, which can be harnessed for medicinal purposes. For example, plants growing at high altitudes are exposed to high UV radiation and low oxygen levels, prompting the production of novel compounds with potent antioxidant, anti-inflammatory, and neuroprotective properties. These molecules hold significant potential for treating diseases like cancer, neurodegenerative disorders, and chronic inflammatory conditions, which currently have limited therapeutic options.

Nanotechnology plays a crucial role in transforming these plant-derived molecules into effective mainstream medicines. By enabling precise delivery, enhanced bioavailability, and targeted action at cellular and molecular levels, nanotechnology overcomes many traditional barriers related to pharmacokinetics and toxicity. For instance, nano-encapsulation can improve the solubility and stability of delicate plant compounds, allowing for controlled release and improved efficacy. This technological advancement also opens the possibility of personalised medicine, where plant-based nanomedicines can be tailored to individual patients' genetic and biochemical profiles.

In essence, leveraging plant molecules from extreme environments combined with cutting-edge nanotechnological approaches promises to revolutionise the treatment landscape, offering new hope for diseases that are currently deemed incurable. This synergy not only accelerates drug discovery but also paves the way for safer, more effective, and accessible therapies, ultimately transforming modern medicine.

•••

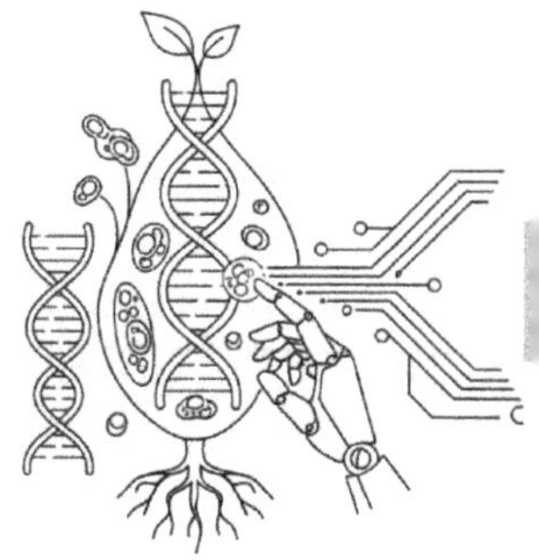

5

HARVESTING POSSIBILITIES

Prosperity is not inherited; it is invented — in the courage to turn knowledge into livelihoods and vision into collective wealth.

Agriculture represents the earliest and most significant example of humans breaking new paths in their way of life. During the transition from hunter-gatherer societies to settled communities, humans began to domesticate plants and animals, transforming their interactions with the environment. This shift allowed for more stable food sources and enabled population growth, leading to the development of villages and civilisations. By cultivating crops and selectively breeding animals, early humans took deliberate steps to modify their surroundings to better suit their needs, marking a pivotal point in human history. This innovation laid the foundation for complex societies, technological progress, and the modern world's diverse agricultural systems.

The sight of high-altitude plants vanishing was deeply painful, especially because these plants held within them the secrets of life—unique adaptations and delicate ecosystems that had awaited discovery and understanding. As development encroached upon these pristine environments, it began with paved roads carving through the landscape, followed by vehicles traversing the onceuntraveled terrain. Trade and commerce soon arrived, bringing with them the allure of economic growth, while tourists flocked to experience the natural beauty. Unfortunately, this influx often shifted the mindset from one of preservation and respect for nature to one of consumption and exploitation.

The delicate balance of the ecosystem was disrupted, and what was once a pure, untouched haven became marred by human greed and short-sightedness. Specifically, in the Lahaul-Spiti Valley, the once-pristine paradise was lost to the construction of a thirsty station, a symbol of development that prioritised immediate needs over environmental stewardship. This transformation highlights the tragic toll of unchecked progress on fragile high-altitude ecosystems and the importance of safeguarding such precious landscapes before they are irreparably damaged. Then COVID-19 arrived.

Various theories were circulated about the origin of the deadly pandemic. I saw it as an encroachment on biodiversity because the pandemic highlighted how human activities and disruptions to natural ecosystems can increase the risk of zoonotic diseases—those transmitted from animals to humans. Habitat destruction, deforestation, and urban expansion lead to closer contact between humans and wildlife, creating opportunities for viruses like COVID-19 to jump from animals (such as bats or pangolins) to humans. China is just a few steps ahead of India when it comes to expanding settlements into pristine ecosystems.

Additionally, the pandemic exposed how overexploitation of natural resources, illegal wildlife trade, and habitat fragmentation weaken ecosystems' ability to maintain ecological balance. These disturbances can lead to increased disease transmission, threatening both human health and biodiversity. Preserving natural habitats and maintaining ecological integrity can reduce the chances of future pandemics and support overall planetary health.

I spent several years working to obtain the seeds of *Heeng* (asafoetida). The dried gum secreted by the tap root of certain species of plants in the *Ferula* genus, part of the carrot family is an essential ingredient in Indian kitchens. It is widely used, especially in vegetarian dishes, lentils, and spice blends, to enhance flavour and aid digestion. *Heeng*'s strong aroma and distinctive taste make it a key component in tempering spices (*Tadka*) and preparing flavourful curries, dals, and pickles. But despite its status as an integral part of Indian kitchens, the entire supply of *Heeng* is imported.

These plants naturally grow in the Karakum Desert, spanning Eastern Iran, Turkmenistan, and the mountainous regions of Afghanistan. After planting a seed, it takes about five years for the plant to mature and yield resin. The stems are cut near the root, releasing a milky sap that quickly hardens into a reddish-brown, resinous mass. Once this resin is harvested, the plant dries out permanently.

Modern scientific analysis has revealed that asafoetida contains volatile sulphur compounds responsible for various biological activities when consumed. Specifically, three major sulphur compounds—2-butyl 1-propenyl disulphide, 1-(methyl thio) propyl 1-propenyl disulphide, and 2-butyl 3-(methyl thio)-2-propenyl disulphide—are consistently found in medicinal products renowned for antimicrobial, hepatoprotective, and anticancer properties.

The supply of *Heeng* faced significant challenges during the prolonged Afghan war starting in 1979 and ongoing militancy in Kashmir, making it exceedingly expensive. India imports approximately 1,200-1,500 tons annually, costing over ₹1,000 crore. Overharvesting from wild populations and the absence of organised cultivation led Iran to declare asafoetida an endangered species.

Belonging to the genus *Ferula*, of the family Apiaceae, of mostly aromatic flowering plants, *F. assafoetida* plant is globally represented by about 170 species and is distributed from the Mediterranean region to Central Asia and in tropical African mountains. *Heeng* is indeed the dried resin of the asafoetida plant, which naturally grows in cold desert regions of Iran, Afghanistan, and Uzbekistan. The four plants of genus *Ferula* are found in India – *F. jaeschkeana* Vatke, *F. narthex* Boiss., *F. thomsonii* C.B. Clarke, and *F. ovina* L. and their gums are used as a cheap substitute for *Heeng*. So, why can the asafoetida be grown in India? I reasoned and assumed the challenge.

Though the thought initiated in 2015, we could obtain the seeds of *F. assafoetida* from Iran and Afghanistan through ICAR-National Bureau of Plant Genetic Resources only in 2018, after a prolonged correspondence and overcoming regulatory hurdles. No one knew how and when these hurdles were created, but still they were jealously guarding them, even more jealously than Angel Rizwan at the gates of heaven, I felt.

Normally, only about 1 in 100 asafoetida seeds germinate under natural conditions, but through careful planning and cultivation techniques, the CSIR-IHBT team achieved a germination rate of approximately two plants for every three seeds. Gas chromatography analysis of a one-year-old plant confirmed the presence of all key sulphur compounds.

The tissue culture laboratory at CSIR-IHBT has developed an efficient regeneration method for asafoetida via somatic embryogenesis. Scientists have successfully cultured cellular masses from roots, leaves, and stems of the mother plant. The high success rate of callus regeneration paves the way for further techniques such as protoplast culture, somatic hybridisation, and genetic transformation. The seeds were germinated with 70% success and I planted the first sapling along with my team in Kwaring village near Keylong in the Lahaul Valley on Dr A.P.J. Abdul Kalam's birth anniversary, October 15, 2020. By this time, Dr Brij Lal was superannuated, and Dr Sanat Sujat Singh and Dr Ashok Kumar were leading the team.

The news was broadcast by the All India Radio while I was on my way back to Palampur. We were having tea at a roadside stall when we heard it on the radio. Within an hour, Dr Shekhar C.

Mande, Director-General, CSIR, tweeted the news. Later, Union Minister for Women and Child Development Smriti Irani called Dr Mande, and he cross-checked with Dr Trilochan Mohapatra, the Director-General of ICAR as well, to confirm whether *Heeng* is grown in India for the first time. By the time I reached home, it was as if an awakening had taken place. What a pity that for centuries, India has been relying on imports due to the lack of local cultivation knowledge and suitable conditions.

I was flooded with requests for supplying *"Swadeshi Heeng"* and struggled to explain that the *F. assafoetida* plant typically takes about 5 years to mature enough to yield significant quantities of resin. Once mature, the roots and stems are incised to produce the exudate, which is then collected as asafoetida. This resin contains volatile compounds and sulphurous compounds that give asafoetida its strong smell and flavour. It is often aged for a few months to develop its characteristic aroma and flavour. Now (in 2025), all this has happened. Our collaborations with the Himachal Pradesh Department of Agriculture led to the cultivation of asafoetida across five districts, with further plantations established in Uttarakhand, Ladakh, and Jammu & Kashmir.

India, renowned for its rich and diverse biodiversity, faces significant challenges in preserving its natural heritage. Despite being home to a vast array of flora and fauna, the country often appears callous in protecting its biodiversity, especially its medicinal plants. This neglect can be attributed to several socio-economic and administrative factors that hinder effective conservation efforts.

One primary reason for India's inadequate preservation of biodiversity is rapid urbanisation and industrialisation. As cities expand and industries develop, natural habitats are often sacrificed for infrastructural projects. Forests, grasslands, and wetlands are being cleared at an alarming rate, leading to habitat destruction and loss of biodiversity. This relentless encroachment diminishes the natural habitats of many medicinal plants, pushing some toward extinction.

Moreover, overharvesting and illegal trade of medicinal plants pose grave threats. Many valuable plants used in traditional medicine, like *Sarpagandha* (*Rauvolfia serpentina* (L.) Benth. ex Kurz), *Ashwagandha* (*Withania somnifera* (L.) Dunal), and various species of Orchids, are collected excessively from the wild to meet high demand. This unregulated exploitation leads to the depletion of these species, making their preservation difficult. Despite the cultural significance of Ayurveda and traditional Indian medicine, systematic cultivation and sustainable harvesting practices are often lacking.

The glamourisation of modern medicine and lack of awareness also contribute to the neglect of medicinal plant conservation. A significant portion of the population prefers allopathic medicine over traditional herbs, leading to decreased demand for these plants. Consequently, there is little incentive for farmers to cultivate medicinal plants on a large scale, causing many of these species to become rare or extinct.

However, it is essential to acknowledge that the Indian government and various organisations have initiated efforts to conserve medicinal plants. Several medicinal plant gardens, conservation projects, and research initiatives aim to protect and cultivate these species. Programs like the National Mission on Medicinal Plants and the establishment of herbal parks illustrate a growing awareness of the importance of conserving biodiversity. Additionally, cultivation of medicinal plants is increasingly encouraged among farmers, and awareness campaigns are being conducted to promote sustainable harvesting.

Despite these efforts, challenges remain. There is still a significant gap between policy implementation and actual practice on the ground. Conservation requires coordinated efforts, stricter enforcement of laws against illegal trade, community participation, and promotion of sustainable cultivation techniques.

While India possesses immense biodiversity and has made some efforts to conserve its medicinal plants, the overall approach often remains callous and insufficient. Addressing this issue demands greater commitment, awareness, and sustainable practices to ensure that India's rich natural heritage is preserved for future generations. Only through concerted efforts can India protect its biodiversity, prevent the extinction of treasured medicinal plants, and continue to benefit from its natural resources.

Tissue culture technology, also known as micropropagation, is a highly effective tool for conserving endangered plants and ensuring their purity. This technique involves growing plant cells, tissues, or organs in a sterile laboratory environment to produce new plants on a large scale. One of the greatest advantages of tissue culture is its ability to rapidly multiply rare and vital plants from a small piece of tissue, making it possible to conserve species that are on the brink of extinction. Additionally, tissue culture can preserve the genetic material of plants, allowing for the storage of their valuable traits *in vitro* for future reintroduction or research.

Tissue culture also helps produce disease-free, genetically identical plants, thus maintaining the purity and quality of medicinal and rare species. By growing tissues in controlled environments, scientists can select and propagate plants that are free from diseases, viruses, or genetic abnormalities, ensuring high-quality, pure plants. This allows for the selective propagation of healthy, pest-free plants, which is crucial for protecting valuable biodiversity. The technique also allows for long-term preservation of plant tissues at very low temperatures, safeguarding the genetic diversity and purity of rare species for future reintroduction or research.

Saffron (*Crocus sativus* L.) is one of the costliest spices, having high demand in India and the world. Saffron cultivation is confined to Kashmir, largely in the Pampore region and produces barely ten tonnes annually against a national demand of almost 100 tonnes. So, we decided to make them *in vitro*, the scientific term for growing cells or organisms outside a living organism in a test tube, culture dish, or other controlled environment.

The tissue culture method allowed us to produce healthy, uniform saffron corms on a large scale. The technique not only increased the availability of saffron but also ensured the production of disease-free corms. By growing saffron corms through tissue culture, India could promote saffron cultivation beyond Kashmir, illustrating the significant role of biotechnology in modern horticulture.

In vitro cultivation of saffron involves growing saffron plants under controlled laboratory conditions using tissue culture techniques. The process begins with selecting small tissue samples, such as stigma, corms, or shoot tips, from mature saffron plants. These explants are then sterilised with disinfectants to eliminate any potential pathogens. Once sterilised, they are placed on nutrient-rich culture media, which contain essential minerals, vitamins, and plant growth regulators like cytokinins and auxins to promote callus formation or shoot proliferation. The explants gradually develop into new shoots or calluses, which are sub-cultured onto fresh media to achieve rapid multiplication. Subsequently, the shoots are transferred to rooting media to develop roots, and the resulting plantlets are gradually acclimatised to external environmental conditions before being transplanted to soil. This *in vitro* method enables the rapid propagation of disease-free plants, conserves genetic traits, and produces uniform, high-quality saffron plants, making it particularly valuable for commercial cultivation and genetic preservation.

The most critical step in the *in vitro* cultivation of saffron is the surface sterilisation of the explants. This step is crucial because it ensures the elimination of all microbial contaminants, such as bacteria and fungi, that could otherwise invade the sterile culture environment and compromise the entire experiment or propagation process. Proper sterilisation must be thorough yet gentle enough to avoid damaging the plant tissues, as contamination can lead to culture failure and loss of valuable genetic material. Effective sterilisation techniques, using appropriate disinfectants and exposure times, are essential for successful tissue culture outcomes. Dr Amita Bhattacharya and later Dr Aparna Maitra Pati worked tirelessly to develop a cost-effective corm production technology for the commercial-scale production of saffron

Through niche modelling, using ecological and environmental data to determine the suitable geographic areas and environmental conditions for optimal saffron growth, Dr Amit Kumar in our team identified the best regions suited for its cultivation in Himachal Pradesh and Uttarakhand in the year 2017 and extended the cultivation in farmers' fields through an agreement with the state government. Typically, niche modelling uses Geographic Information Systems (GIS) and algorithms such as MaxEnt or CLIMEX to analyse variables like temperature, rainfall, soil pH, and sunlight. By integrating these data, farmers and researchers can predict potential cultivation zones, optimise crop management practices, and assess the impact of climate change on saffron production. This strategic tool enhances decision-making, increases yield efficiency, and supports sustainable cultivation practices by guiding location-specific intervention measures. The quality and yield of saffron in non-traditional areas were found to be comparable to those produced in Kashmir.

The famed Tulip Garden in Srinagar, Kashmir, has been importing ₹15 crore worth of tulip bulbs (*Tulipa gesneriana* L.) from the Netherlands every year. I felt outraged. I discussed the matter with my colleagues and Dr Bhavya Bhargava decided to produce tulip bulbs indigenously. We acquired tulip bulbs and identified the locations where the bulbs could be propagated. Bulb development needs cooler soils, and it emerged that locations in Lahaul and Ladakh would favour bulb production. We enrolled the farmers there and developed appropriate agrotechnology for bulb production.

When our news circulated through newspapers, the New Delhi Municipal Council and later the officer of the Lieutenant Governor of Delhi contacted us for the indigenisation of tulip bulb production. It gives me a profound sense of satisfaction when the Tulip Garden in Delhi starts taking our material, at least in part. Indian tulips grow side by side with the Dutch ones, and it is very difficult for anyone to differentiate. Even CSIR-IHBT has its own tulip garden, which is open to the public. The best time to enjoy the full bloom of tulips in these regions is usually from late February to April, when the weather is cool and conducive to their growth.

We also pioneered aeroponics and hydroponics, the innovative soil-less cultivation methods used to grow plants, offering efficient resource use and often resulting in faster growth and higher yields. In aeroponics, roots are suspended in the air within a closed environment and are regularly misted or sprayed with a nutrient-rich water solution, providing essential nutrients and oxygen directly to the roots. In hydroponics, the roots are submerged in a growing medium, such as coco coir, perlite, or rockwool, while absorbing nutrients directly from the water. Hydroponic systems can be configured in various ways, including nutrient film technique (NFT), deep water culture, and flood-and-drain systems, enabling efficient control over nutrient delivery, water use, and environmental conditions.

Picrorhiza kurroa Royle ex Benth, commonly known as kutki or Indian gentian, is a perennial herb belonging to the family Plantaginaceae. It is native to the Himalayan regions of India, Nepal, and Tibet, typically growing at high altitudes. The plant is characterised by its short, tuberous roots and purple or violet flowers. The root of *P. kurroa* is highly valued in traditional medicine systems like Ayurveda and Unani for its medicinal properties. It is primarily used as a hepatoprotective agent to treat liver disorders, hepatitis, and jaundice. The plant also exhibits anti-inflammatory, anti-diabetic, and antioxidant properties. Due to its medicinal importance, *P. kurroa* is cultivated and harvested for pharmaceutical purposes, and efforts are underway to conserve its natural populations due to overharvesting and habitat loss. It takes about three years to grow in the field. However, integrating tissue culture and hydroponics and aeroponics Dr Ashish Warghat achieved healthy, rapid growth and a successful, quick harvest of *P. kurroa* in merely 5 months.

Valeriana jatamansi Jones is highly valued in traditional medicine systems. The rhizome and roots are used to treat a variety of ailments, including mental health issues like anxiety, insomnia, and neurological disorders, owing to their calming and sedative effects. It also has anti-inflammatory, antimicrobial, antioxidant, and skin-healing properties. Additionally, the fragrant rhizome is used in perfumery and aromatherapy. Conserving their natural populations due to overharvesting has become a serious issue. We achieved growing it aeroponically in 4 months as opposed to 2 years in the field.

Ocimum basilicum L., commonly known as sweet basil, is widely used in cooking, particularly in Italian, Mediterranean, and Southeast Asian cuisines, to flavour dishes such as pesto, salads, soups, and sauces. Apart from culinary uses, basil has medicinal properties; it is rich in antioxidants and has anti-inflammatory, antimicrobial, and digestive benefits. Traditionally, basil has been used to treat ailments such as indigestion, respiratory issues, and stress. Additionally, basil essential oil is used in aromatherapy and cosmetics. Due to its versatile applications, basil is cultivated both commercially and in home gardens worldwide. We grew it hydroponically in six months' time. Tulips are also now grown hydroponically, and flowers can be obtained at any time of the year. This technology enabled us to offer tulip flowers to Lord Shri Ram during the Pran Pratishtha ceremony on January 22, 2024, which would have otherwise been outside the natural tulip flowering season.

We broke a new path in fermentation technology or bioreactor cultivation for culturing plant cells and tissues, which involves utilising controlled bioreactor systems to grow plant cells, tissues, or organs *in vitro* on a large scale. This approach enables the production of valuable active components, such as secondary metabolites, pharmaceuticals, and bioactive compounds, in a cost-effective and scalable manner. The process typically includes selecting appropriate explants, establishing sterile culture conditions, and optimising growth media with specific nutrients, plant growth regulators, and elicitors. Parameters such as pH, temperature, aeration, and agitation are carefully regulated within bioreactors to maximise cell proliferation and metabolite accumulation. Fermentation of plant cells offers advantages like high yield, consistency, and the ability to produce rare or difficult-to-obtain compounds without the need for entire plants, making it a powerful tool for pharmaceutical, cosmetic, and nutraceutical industries.

Arnebia euchroma (Royle ex Benth.) I.M. Johnst., commonly known as Dyer's Arnebia, is highly valued for its medicinal and economic importance. The roots contain principal active compounds such as shikonin and its derivatives such as acetylshikonin, which are renowned for their pharmacological properties. They are used in traditional medicine systems to treat various ailments, including wounds, infections, and skin conditions. Additionally, the root extracts are used as natural dyes, pigments, and in cosmetic formulations. Due to overharvesting and habitat loss, conservation efforts became rather urgent.

In vitro production of *A. euchroma* involves cultivating the plant's tissues, such as callus, roots, or cell suspensions, under sterile, controlled laboratory conditions to generate shikonin. The process begins with selecting and sterilising explants from the plant, then inducing callus formation using specific plant growth regulators in nutrient-rich media. From this callus or cell suspension, secondary metabolites are produced through optimised culture conditions, which may include elicitors or precursor feeding to enhance compound synthesis. These *in vitro* cultures enable mass production of active ingredients without the need for wild harvesting, ensuring a sustainable supply of pharmacologically important compounds while conserving natural populations of *A. euchroma*. What takes five years in the field, fetching ₹5,000 per kg of dry roots, we achieved in four months *in vitro*. I had nominated one of my scientists, Dr Shashi Bhushan, to learn this technique and this helped our institute to conserve the endangered species *A. euchroma* for sustainable production and minimise the field harvesting.

Yet another new path was shiitake mushroom (*Lentinula edodes* (Berk.) Pegler). It is a popular edible fungus widely cultivated for its rich flavour, meaty texture, and nutritional benefits. Characterised by their brown caps, white stems, and gills on the underside of the cap, shiitake mushrooms are highly valued in culinary dishes around the world for their savoury umami taste. Beyond their culinary appeal, they are known for their health benefits, which include boosting the immune system, possessing anti-inflammatory and antioxidant properties, and potentially reducing cholesterol levels.

Fresh shiitake mushrooms typically contain very low levels of vitamin D_2, usually less than 1 microgram per 100 grams. However, when exposed to UV light, the vitamin D_2 content can increase dramatically to 50 micrograms per 100 grams or even higher, making it a good dietary source of vitamin D_2. Dr Rakshak Kumar developed a cultivation technology to enrich vitamin D_2 content by stimulating the conversion of ergosterol, a naturally occurring compound in the mushroom cell membranes, into vitamin D_2. He exposed the mushrooms to UV-B light for a specific duration (30 minutes to 2 hours) in a controlled environment and then quickly dried them. Regular consumption of UV-treated shiitake mushrooms can contribute meaningfully to daily vitamin D intake, supporting bone health and immune function.

The first part of the book has now come to an end. In the five chapters starting with a pollen grain riding apparently random wings, I became a plant physiologist and landed up at the CSIR-IHBT to work on high-altitude bioresources. I shared how the quest for understanding the secrets behind phenomena fuels a passion. In the cold desert of Spiti Valley, I articulated how plants survive the harshest possible climate. If you remove a thick layer of snow, you can find a tender plant waiting to grow. I discovered the metabolites, special proteins responsible for this survival, as a new area of research that eventually merged into the larger field of ecobiology, an interdisciplinary field that combines principles of ecology and biology to study the relationships between living organisms and their environment. This gave birth to the Phytopharmaceutical, Aroma, Floriculture and Nutraceutical missions. Our mastery of tissue culture led to a successful enterprise of growing high-value plants in an accelerated and controlled manner.

In contemporary times, the concept of a bioresource-based economy or bioeconomy has gained immense importance as a sustainable and innovative approach to harnessing natural resources. It emphasises transforming high-quality biological assets such as aromatic and medicinal plants, fruits, spices, and bamboo from traditional, small-scale practices into large-scale, value-added sectors. This transition involves not only increasing the quantity of production but also maintaining or enhancing the quality of these products, thereby creating a balanced ecosystem that benefits farmers, industries, and the environment.

Aromatic and medicinal plants hold a pivotal place in the bioeconomy, owing to their extensive use in pharmaceuticals, cosmetics, and nutraceuticals. Traditional methods, often limited to subsistence farming, are now being upscaled through scientific cultivation, tissue culture, and biotechnological interventions. By standardising cultivation practices, employing high-yielding and disease-resistant varieties, and adopting biotechnological tools like genetic improvement and tissue culture, it is possible to boost both the quality and quantity of these plants. For example, plants like *Ashwagandha* [*Withania somnifera* (L.) Dunal], basil (*Ocimum basilicum* L.), and sage (*Salvia officinalis* L.) are being cultivated with enhanced phytochemical profiles in large plantations. Through value addition, extraction, and processing, the high-quality raw materials are translated into multiple economic products, thus fostering a vibrant bioeconomy.

Fruits and spices are integral to the food, flavour, and health industries. The core challenge lies in translating their intrinsic quality—rich flavours, aroma, and nutritional content into large-scale production. Modern horticultural practices, organic farming, integrated pest management, and precision agriculture enable the scaling of fruit and spice cultivation without compromising their quality. For instance, spices like saffron (*Crocus sativus* L.), turmeric (*Curcuma longa* L.), and pepper (*Piper nigrum* L.), and fruits like mango (*Mangifera indica* L.) and pomegranate (*Punica granatum* L.) are now cultivated under controlled conditions to enhance their bioactive compounds and sensory attributes. Additionally, upscaling involves establishing processing units for drying, grinding, and extracting essential oils that retain the high-quality features of raw materials while ensuring larger availability. This approach not only fulfils domestic demand but also boosts exports, creating employment and economic growth.

Bamboo, often termed as 'green gold,' exemplifies a renewable resource with vast potential. It has applications across construction, furniture, paper, textiles, and even the food industry. To translate bamboo's quality into quantity, advanced cultivation methods, including clonal propagation, standardised nurseries, and integrated agroforestry, are being employed. These techniques ensure the cultivation of high-quality bamboo with desirable strength and fibre properties at larger scales. Furthermore, value addition through processing—such as converting bamboo into engineered products, bamboo charcoal, or bioenergy—further amplifies its economic impact. Bamboo's sustainable growth and versatile uses make it a cornerstone of the bioeconomy.

The central premise of a bioeconomy is to firmly establish that scaling up does not mean compromising quality. It requires integrating traditional knowledge with modern science, ensuring sustainable harvesting, and deploying innovative technologies.

Policies promoting organic farming, certification standards, and value chain development are fundamental. Additionally, fostering research and development—focusing on bioactive compound extraction, cultivation techniques, and product diversification—will bridge the gap between quality and quantity.

Now, I will take you to Part 2 of the book, where we unveil the immense potential of the bioeconomy, based on aromatic and medicinal plants, fruits, spices, and bamboo, for sustainable development, rural entrepreneurship, and global trade. By translating the inherent quality of these resources into increased quantity through scientific, technological, and managerial interventions, we can create a resilient and inclusive bioeconomy. This paradigm shift not only conserves biodiversity and ecological balance but also ensures that the benefits gleaned from nature reach the masses, fostering a healthier planet and prosperous communities.

•••

PART II
THE PATH AHEAD

"Technology is a non-linear tool. It can affect fundamental changes in the ground rules of economic competitiveness."

Dr A.P.J. Abdul Kalam
(1931 – 2015)

6

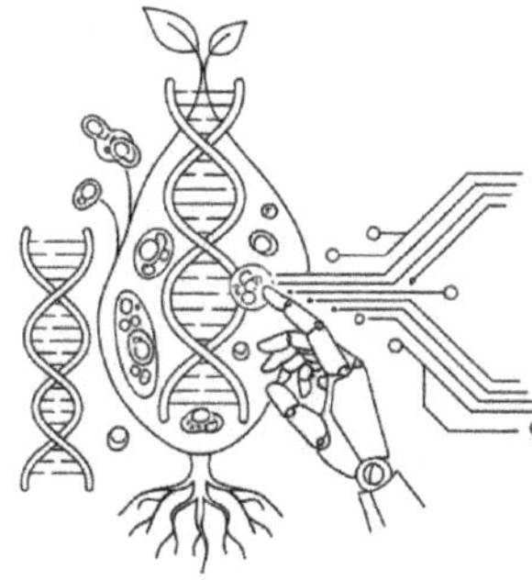

BLOOMING PROSPERITY

India's floral wealth is not merely a harvest of petals,
but a reservoir of destiny. By fusing science with nature's
abundance, we can craft an aroma legacy
that elevates our nation's spirit and stature across the world.

I start the second part of the book with the economic impact of flowers and perfumes. Interestingly, many perfumes are made from essential oils and extracts derived from flowers and are often considered "children of flowers", they take precedence over flowers because they offer lasting scent, controlled application, and versatility. Unlike flowers, which are short-lived, perfumes can be easily carried and stored for long periods. Perfumes can be worn anytime and anywhere, providing a consistent fragrance experience, whereas flowers emit their scent for a limited time and are typically used for decoration or natural ambience.

France is renowned for its significant perfume industry, which plays a vital role in its economy and global cultural influence. The perfume industry in France dates back centuries, with Grasse historically serving as the centre of perfume production due to its ideal climate for growing fragrant flowers like jasmine (*Jasminum sambac* (L.) Aiton), rose (*Rosa* × *damascena* Mill.), and violet (*Viola odorata* L.). France is home to numerous prestigious perfume houses such as Chanel, Dior, L'Oréal, Guerlain, and Yves Saint Laurent. These brands contribute significantly to France's exports and economy, generating billions of euros annually. It provides numerous jobs in manufacturing, marketing, research, and retail.

The Provence-Alpes-Côte d'Azur (PACA) region of southeastern France is a stunning and diversified area known for its beautiful landscapes, rich history, and vibrant culture. Situated in the Côte d'Azur, also called the French Riviera, Grasse has been a centre for the cultivation of fragrant flowers such as jasmine, rose, lavender, and violet. These flowers are essential raw materials for perfume production. The town hosts numerous perfume factories, workshops, and museums dedicated to the art of fragrance making. Many top perfume houses have origins or production facilities in Grasse. Visitors to Grasse can tour perfume museums, participate in scent-making workshops, and explore the picturesque old town, which is filled with perfumeries and botanical gardens.

Guerlain is one of the oldest and most prestigious perfume and cosmetics houses in the world, with a history that dates back to 1828. The brand was founded by Pierre-François Pascal Guerlain in Paris. Originally established as a perfumery and pharmacy, Guerlain quickly earned a reputation for its high-quality fragrances and innovative formulations. Guerlain became renowned for creating some of the most iconic and luxurious perfumes in the world, like Shalimar in 1925. Guerlain was also innovative in its approach to packaging, blending art and science to create beautifully designed bottles and containers that enhanced the luxury experience.

In 1994, Guerlain was acquired by the Louis Vuitton Moët Hennessy (LVMH) group, which helped further solidify its status as a global luxury brand. Under LVMH's stewardship, Guerlain has continued to innovate while preserving its rich heritage, combining traditional craftsmanship with modern technology. Today, Guerlain remains a symbol of French luxury and elegance, celebrated for its pioneering fragrances, high-quality products, and timeless appeal, influencing the perfume and beauty industry worldwide.

Chanel is one of the most iconic and influential luxury brands in the world. In 1910, Gabrielle "Coco" Chanel opened a millinery shop, offering stylish hats in Paris. Her vision was to create clothing that emphasised simplicity, comfort, and elegance, which was revolutionary in an era dominated by restrictive and elaborate fashion and popularised the use of comfortable fabrics like jersey for chic clothing. The brand's iconic interlocking "CC" logo was introduced in the 1920s. One of Chanel's most famous products, Chanel No. 5 perfume, launched in 1921, remains one of the best-selling and most recognisable perfumes worldwide even after 100 years. A complex blend of floral and aldehyde notes indeed set a new standard in fragrance.

Dior was established in 1946 by French designer Christian Dior in Paris. Lifting the mood of the Second World War, where French people suffered German military occupation and humiliation, Dior introduced the "New Look." It featured cinched waists, full skirts, and an overall emphasis on femininity and elegance, transforming the fashion landscape and symbolising a return to opulence after years of austerity. Christian Dior continued to innovate in fashion, expanding his empire to include accessories and perfumes. The iconic Miss Dior perfume was launched in 1947. After Christian Dior's death in 1957, the house continued to evolve with designers like Yves Saint Laurent, Gianfranco Ferré, John Galliano, and later, Kim Jones, each bringing their own vision while preserving Dior's legacy of elegance and innovation.

L'Oréal was founded in 1909 by Eugène Schueller, a French chemist. Schueller developed a revolutionary hair dye formula, which he initially sold to Parisian hairdressers. His innovative approach to hair colouring and beauty products laid the foundation for what would become a global cosmetics empire. Throughout the early 20th century, L'Oréal expanded its product range and international presence. The company introduced a variety of beauty products, including shampoos, skin care, and makeup, and became known for its emphasis on scientific research and innovation. L'Oréal operates in over 150 countries and owns a broad portfolio of brands across various beauty sectors, including luxury, professional, and consumer products.

Yves Saint Laurent, often abbreviated as YSL, was founded in 1961 by designer Yves Saint Laurent and his partner, Pierre Bergé, in Paris. From its inception, YSL revolutionised fashion by introducing ready-to-wear collections that combined high fashion with accessibility, breaking away from the traditional couture model. Their tuxedo jacket for women, known as "Le Smoking," challenged gender norms and became an iconic symbol of modern femininity and empowerment. Its debut perfume, Y, was launched in 1964. YSL remains a symbol of modern elegance, creativity, and rebellion in fashion, transforming the landscape of fashion and beauty for over six decades. Shumukh by Nabeel is considered amongst the world's most expensive perfumes.

Chanel at $18-20 billion in market capitalisation (in 2023) is the leader, followed by Dior, L'Oréal, YSL, and Guerlain, which is valued at $7.7 billion in 2025. To put things in perspective, though it is a privately held company, its exact market valuation is not disclosed, but Patanjali's revenue could be in the range of $3 billion to $5 billion at different points.

Kannauj in Uttar Pradesh is our "Perfume Capital." It has a rich tradition of making *Attar,* oil-based perfumes. Essential oils are obtained from flowers, herbs, and woody resins and then diluted with a carrier oil like sandalwood oil or coconut oil. The aroma of attar is developed through oxidation, ageing, and polymerisation processes, which convert precursors into volatile aromatic compounds. Enzymatic reactions during distillation or aging can produce aldehydes, phenols, and other aromatic compounds that define the final scent. The city's perfume industry has a unique, artisanal approach that emphasises craftsmanship, with many small, family-run workshops continuing age-old techniques. Nearby Firozabad also makes perfumes which are often more intense, bold, and sometimes modern in their aroma compositions.

The CSIR-CIMAP, Lucknow, has established and standardised major components in *Attar*. Terpenoids, also called isoprenoids, are found throughout the natural world, including in plants, insects, and even some microorganisms and are responsible for fragrance notes. Phenylpropanoids like eugenol, vanillin, and safrole are synthesised by plants. These are characterised by a three-carbon (C3) side chain attached to a benzene ring (C6), making the core C6-C3 structure. Aromatic aldehydes, like benzaldehyde, give an almond-like odour, and coumarin gives a sweet, vanilla-like aroma. The biochemical pathways involved are also well understood. The mevalonate pathway produces isoprenoid precursors. It leads to the synthesis of terpenoids, which define many of the characteristic scents. The shikimate pathway produces phenylalanine, a precursor for phenylpropanoids and aromatic aldehydes, which contribute to the scent and flavour profiles.

Coming to flowers, the Netherlands is world-famous for its tulip industry. The Dutch tulip industry played a central role in what is known as "Tulip Mania." During the Dutch Golden Age, around 1637, tulip bulbs became highly fashionable and their prices soared to extraordinary levels, with some bulbs costing more than a skilled tradesman's house. People bought and sold tulip bulb contracts as a form of investment, often on credit, expecting prices to keep rising. Ultimately, the market collapsed, resulting in substantial financial losses for many investors. Since then, Tulip Mania is cited as an example of economic bubbles driven by speculative frenzy and irrational exuberance.

Nevertheless, the Netherlands remains the global leader in tulip cultivation, production, and export. The country is home to vast fields of tulip gardens, especially in regions like Keukenhof, which is one of the world's largest flower gardens and a major tourist attraction. The Dutch tulip industry is renowned worldwide for its high level of organisation and sophistication. It encompasses a comprehensive and well-structured supply chain that includes various specialised professionals. Bulb breeders focus on developing new, improved tulip varieties with desirable traits such as vibrant colours, disease resistance, and longer-lasting blooms. Cultivation experts and horticulturists manage large-scale farms, employing advanced techniques and technology to optimise growth conditions, ensure quality, and increase yield. The industry also involves efficient logistics and export professionals who handle the transportation, packaging, and international distribution of tulip bulbs and flowers, maintaining strict quality standards. This integrated approach has established the Netherlands as the global leader in tulip production, contributing significantly to both the economy and cultural heritage.

Bulgaria has a highly developed flower industry, renowned particularly for its production of roses, lavender, and other aromatic plants. Historically, the cultivation of roses in Bulgaria dates back to ancient Thracian times, around the 2nd millennium BC, when Indo-European tribes inhabited regions of Southeast Europe, particularly in areas that are now Bulgaria, Greece, and Turkey. The tradition flourished during the Ottoman period, from the late 14th century, when a powerful Islamic monarchy ruled over vast territories in Southeast Europe, Western Asia, and North Africa. At this time, this region became a centre for producing rose oil, a luxury commodity used in perfumery, cosmetics, and medicinal products. The climate, soil, and traditional distillation techniques contributed to the unique aroma and high purity of Bulgarian rose oil, making it highly sought after globally.

The modern concept of the Bulgarian Rose Valley as a dedicated region for rose cultivation and essential oil extraction was officially recognised in the 20th century. Today, the Valley hosts an annual Rose Festival in June, celebrating the rose harvest, and attracts thousands of tourists and industry professionals. It remains a vital hub for the global perfume industry, exporting large quantities of Bulgarian rose oil, often called "Liquid Gold," which is a cornerstone ingredient in many luxury fragrances. In addition to roses, Bulgaria also cultivates lavender and other aromatic herbs used in the fragrance industry, as well as cut flowers for domestic and export markets. The flower industry in Bulgaria supports numerous small-scale farmers and large agricultural enterprises, contributing significantly to local economies and exports.

Apart from the Netherlands and Bulgaria, several other countries are major players in the global flower industry, known for their large-scale cultivation, export, and innovation in floriculture. Colombia in South America is known as one of the world's largest exporters of cut flowers, particularly roses, carnations, and lilies. Neighbouring Ecuador grows high-quality roses and exotic flowers. Its mountainous terrain and ideal climate support year-round flowering. Kenya in East Africa is a major exporter of roses and flowers to Europe. Kenya's large flower farms benefit from a suitable climate and efficient transportation infrastructure. The Kenyan flower industry has grown significantly over the past few decades. Neighbouring Ethiopia is emerging as a key player in flower exports. Ethiopia's favourable altitude and climate have attracted numerous international flower farms.

Our own Valley of Flowers in the Chamoli district of Uttarakhand is a renowned national park and UNESCO World Heritage Site. It is famous for its stunningly diverse and vibrant alpine flora, a term for high mountain regions, especially those resembling the Alps or mountainous areas above the tree line. The Valley of Flowers is situated at an altitude of about 12,000 feet, spanning over 85 square kilometres. Rare species like the blue poppy [*Meconopsis aculeata* Royle (Himalayan Blue Poppy, most likely species in Valley of Flowers)], cobra Lily (*Arisaema propinquum* Schott), and primulas (*Primula denticulate* Sm.) blossom here. It is home to diverse wildlife, including rare species like the Himalayan Monal, Asiatic black bears, and musk deer. The Valley also has mythological and spiritual importance, being linked to local legends and worship practices of the region.

While "Valley of Flowers" is more famously associated with Uttarakhand, Jwara in Hampta Pass trek in Kullu district of Himachal Pradesh, a corridor between the Kullu Valley and the Lahaul Valley is also known as the "Valley of Flowers". This is known for its spectacular upland meadows and dazzling wildflowers such as Himalayan blue poppy, yellow marigolds (*Tagetes minuta* L.), and pink hill geranium (*Geranium wallichianum* D. Don ex Sweet).

The place also unveils an unforgettable experience of a flower-filled paradise.

India's diverse climatic zones support a wide variety of flowers, making the country a significant producer and exporter of floral plants and cut flowers, with numerous local hotspots catering to both domestic and international markets. Kashmir is a famous flower-growing region, especially for saffron, which is one of the most valuable spices and flowers in the world. The region also produces a variety of ornamental and flowering plants. Himachal Pradesh and Uttarakhand states are known for their diverse flora, including orchids, lilies, and other seasonal flowers. The cooler climate supports the cultivation of various ornamental and medicinal plants.

Ooty and Kodaikanal in Tamil Nadu are famous for their vibrant flower gardens and nurseries, featuring a variety of roses, azaleas, marigolds, and chrysanthemums. Pune and nearby regions in Maharashtra have flower nurseries producing roses, marigolds, and chrysanthemums. Punjab and Haryana grow marigolds and jasmine. Chittoor and Nellore districts in Andhra Pradesh are known for hibiscus and jasmine cultivation. Nadia, Murshidabad in West Bengal and Assam are known for their beautiful lotus [*Nelumbo nucifera* (Gaertn.)] and water lily cultivation.

The lotus (genus *Nelumbo*) is India's national flower because of its deep cultural, spiritual, and historical significance. It's a symbol of purity, enlightenment, and rebirth in Indian culture, and it also represents resilience and the triumph of good over evil. The flower's ability to bloom beautifully despite growing in muddy water is a powerful metaphor for overcoming adversity. The science behind this involves a combination of surface structure, water-repellent properties, and self-cleaning mechanisms. Its botanical features demonstrate adaptations to aquatic life, including specialised reproductive and structural traits. Encased in large, woody seed heads that can remain dormant for years. Seeds are aquatic-dispersed and capable of prolonged viability.

The lotus leaf's surface is covered with microscopic wax-coated bumps, creating a rough texture that minimises the contact area between water and the leaf. The waxy coating on the surface makes it highly water-repellent, causing water to bead up into nearly spherical droplets. As water droplets roll off the leaf, they pick up dirt, microbes, and other debris. This self-cleaning mechanism is called the "lotus effect" and helps keep the leaf clean and dry. The waxy layer on the leaf's surface enhances hydrophobicity, providing the lotus with its iconic ability to remain clean even in muddy environments, which has inspired biomimetic designs for waterproof surfaces, self-cleaning materials, and anti-fouling coatings.

But with all that hype, adulation, veneration, and celebration, India's annual flower exports have rarely crossed $100 million. So, what do we do to enhance the Indian flower and perfume industry by modernising cultivation practices? The government recognised the strategic importance of aromatic plants for rural livelihoods and export earnings. It provided funding, subsidies, and policy support to encourage cultivation, research, and industry development. The Aroma Mission was launched by CSIR as a comprehensive initiative aimed at promoting the cultivation of aromatic plants, improving the quality of essential oils, and boosting India's global presence in the aromatic and perfume industry. It was initiated by the Indian government to harness the economic potential of aromatic plants, which are cultivated for essential oils used in perfumes, cosmetics, and aromatherapy.

Despite its promising objectives and initial momentum, the Aroma Mission has faced several challenges that saw incremental solutions. Most cultivation is done by small-scale farmers without proper organisation or access to markets. While there were efforts to promote R&D, the industry still lags in adopting advanced cultivation and extraction technologies. This limited the ability to innovate and meet international quality standards consistently. But perhaps the biggest impediment is that insufficient processing units and inefficient supply chain infrastructure hinder the rapid scaling of aromatic plant cultivation and essential oil production. This affected product quality and export competitiveness.

There is no panacea. To enhance this industry, a multi-faceted approach focusing on modernisation, quality improvement, infrastructure development, and promoting research and innovation is needed. Firstly, India needs to invest in modern cultivation techniques and research to increase the yield and quality of flowers and aromatic plants. Implementing advanced agricultural practices and introducing high-yield, disease-resistant varieties can help farmers produce better quality flowers, which are essential for the perfume industry. Secondly, establishing well-equipped processing units and packaging facilities is crucial. These units should adhere to international standards to ensure the preservation of fragrance and freshness. Proper packaging can also extend the shelf life of floral and aromatic products, making them more competitive in global markets. Thirdly, the government and private sector should promote research and development to explore new aromatic species and develop innovative fragrances. Collaboration with institutions specialising in botany, chemistry, and fragrance technology can lead to the creation of world-class perfumes and essential oils.

Furthermore, India needs to focus on branding and marketing strategies to promote its floral and perfume products internationally. I have discussed the great French brands with that purpose. Participating in global trade fairs, establishing a dedicated export infrastructure, and obtaining certifications can enhance India's reputation as a reliable source of high-quality floral and aromatic products. In addition, encouraging farmers and entrepreneurs through subsidies, training, and access to modern technology can drive growth in rural areas, ensuring sustainable production and employment generation.

Technology can play a transformative role in facilitating a flower revolution in India by improving productivity, quality, and market access. Advanced agricultural technologies such as precision farming, IoT-based sensors, and drone technology can help farmers monitor soil health, water usage, and crop health with greater accuracy. This enables optimal resource utilisation, reduces wastage, and results in higher yields of high-quality flowers. Implementing smart irrigation and fertilisation systems ensures that flowers receive the right nutrients at the right time, leading to healthier plants and more vibrant blooms.

Innovation in biotechnology, such as the development of hybrid flower varieties through genetic engineering, can produce flowers with improved resilience to pests, diseases, and changing climate conditions. This can extend the flowering season, reduce dependency on chemical pesticides, and ensure a steady supply of premium flowers for markets and exports. Additionally, tools such as tissue culture techniques can help rapidly multiply elite flower varieties, making high-quality planting material more accessible to farmers.

Technology also opens new avenues for marketing and export. E-commerce platforms and mobile apps allow farmers and entrepreneurs to directly connect with buyers worldwide, bypassing middlemen and fetching better prices. Digital platforms can provide real-time market information, price trends, and weather updates, helping farmers make informed decisions about planting and harvesting schedules.

Furthermore, innovative packaging and cold chain logistics driven by technology can help maintain flower freshness during transportation, expanding the reach of Indian flowers to international markets. One can see growers sitting with their baskets of flowers on buses and trucks to sell them in cities. Many sell them at pittance to intermediaries who in turn sell them to cities. Sustainable practices, like solar-powered fertigation systems and eco-friendly cultivation methods, can be integrated through technology to promote environmentally responsible farming.

Back to science, tissue culture is one of the most commonly used biotechnological methods in rose cultivation. It allows for rapid multiplication of elite rose varieties, ensuring the production of disease-free, uniform plants that can be distributed widely to farmers and commercial growers. This method also helps conserve rare or endangered rose species and facilitates the development of new cultivars through somaclonal variation and mutation breeding.

Somaclonal variation refers to the genetic variability that arises in plants regenerated from tissue culture (callus, cell suspension, or shoot cultures). This variation is spontaneous and can lead to desirable traits. By cultivating rose tissues *in vitro*, we can utilize natural genetic variability to develop new flower colours, forms, or improved disease resistance. Variants exhibiting unique colour and shades, increased flower size, or better stress tolerance can be selected and propagated. Mutation breeding involves exposing plant material to physical (gamma rays, X-rays) or chemical mutagens (ethylmethanesulfonate, EMS) to induce genetic mutations that can lead to novel traits. Both these techniques are especially useful in creating ornamental roses that meet evolving consumer preferences and environmental challenges.

Elite rose varieties refer to high-quality, specially bred roses that possess desirable traits such as vibrant colour, unique fragrance, disease resistance, long-lasting blooms, and adaptability to various climatic conditions. These varieties are often developed through advanced breeding techniques, including hybridisation and genetic selection, to meet the demands of both the commercial floral industry and individual growers. Some well-known elite rose varieties include hybrids like *"Sonia,"* and *"Tropicana,"* which are prized for their large, attractive flowers and vibrant colours. These roses are extensively used in cut flower arrangements, landscaping, and decorative purposes due to their aesthetic appeal and robustness. Breeders continuously work to develop new elite varieties that combine aesthetic qualities with resilience against pests, diseases, and environmental stresses.

In addition to their visual appeal, elite varieties often have enhanced fragrance profiles, making them highly desirable for perfume and aroma derivative industries. The development of such varieties involves knowledge of genetics, biotechnology, and traditional breeding practices to create roses that meet specific market needs—whether for ornamental use, event decoration, or commercial production. I consider elite rose varieties represent the pinnacle of horticultural innovation and are crucial for advancing the floriculture industry by offering high-quality, commercially viable, and aesthetically superior options for growers, exporters, and consumers worldwide.

But roses are not merely an ornamental flowers. Roses are known for their anti-inflammatory effects, which can help reduce skin redness and irritation, while their antioxidant properties protect cells from oxidative damage. Rose extracts and rose water are also used to soothe the digestive system, easing stomach discomfort and promoting better digestion. The aroma of roses has a calming effect on the mind, reducing stress and anxiety, making them valuable in aromatherapy. In skincare, rose water and rose oil are appreciated for their hydrating, anti-ageing, and healing properties, beneficial for maintaining healthy skin. Additionally, rose extracts may help alleviate menstrual discomfort due to their mild analgesic effects, and they can soothe respiratory issues like coughs and colds. Overall, roses have a wide range of therapeutic uses, but it is advisable to use rose-based products properly and consult healthcare providers when using them for medicinal purposes.

High-value medicinal rose varieties include *Rosa damascena* Mill. (Damask Rose), *Rosa centifolia* L. (Cabbage Rose or Hundred-Leaf Rose), *Rosa gallica* L. (French Rose), *Rosa roxburghii* Tratt. (Chunpi Rose), and *Rosa multiflora* Thunb, loaded with potent antioxidants and bioactive compounds. These rose varieties are cultivated not only for their beauty but also for their medicinal properties, which are harnessed in herbal remedies, skincare products, and supplements.

I must mention the dhatura (*Datura metel* L.) flower here, which is highly toxic but also of great medicinal value. A large trumpet-shaped flower, often 4 to 12 inches long, with a pleasant fragrance it is found in white, purple, or yellow. All parts of the plant, including flowers, contain tropane alkaloids (atropine, scopolamine, hyoscyamine), which are highly toxic and can cause hallucinations, poisoning, or even death if misused. No wonder, it is offered to Lord Shiva. Yet another tropane alkaloid-containing plant called belladonna (*Atropa belladonna* L.), this plant is found in Europe and North Africa and is popular as Deadly Nightshade.

The metaphor "the elephant in the room" comes to my mind. This metaphor refers to a significant problem, issue, or uncomfortable truth that is obvious to everyone present but is being ignored or deliberately avoided. It highlights the tendency to overlook or avoid addressing a difficult subject despite its clear importance. The elephant in the room here is fragmentary cultivation and the absence of storage and processing at the site. Economic scale and branding are additional factors that influence the growth and global competitiveness of the Indian flower and perfume industry. Not scientific knowledge, not lack of skills or investment, but these deficiencies are among the root causes hindering the industry's full potential to take off.

Most flower and aromatic plant cultivation in India is carried out by small-scale farmers without proper organisation or consolidation. This fragmentation prevents achieving economies of scale, leading to high production costs and inconsistent quality. The lack of large, modern processing and extraction units limits the mass production of essential oils and perfumes. Small units are unable to invest in advanced technologies or meet international standards cost-effectively. Poor logistics and warehousing infrastructure lead to wastage, delays, and quality degradation, further increasing costs and reducing competitiveness in exports.

India has not established a strong, recognisable brand for its floral and aromatic products akin to France's Grasse or Bulgaria's Rose Valley. This limits India's visibility and perceived credibility in the global perfume market. Indian products often lack aggressive branding, global marketing campaigns, and participation in international trade fairs, reducing their market appeal and consumer awareness. Without consistent quality standards and certifications recognised internationally, Indian products face challenges gaining trust and premium pricing abroad.

The inability to produce at scale affects pricing, leading to low profit margins that discourage investments in innovation and quality improvement. Weak branding prevents Indian products from commanding higher export volumes and premium prices in competitive markets. Absence of a strong domestic and international brand restricts market diversification and expansion into premium segments. Without addressing issues related to achieving large-scale production and establishing a strong, recognisable brand, India's flower and perfume industry will struggle to compete globally and realise its full economic potential. Focused efforts on scaling up production, standardisation, and strategic branding are essential for unlocking sustainable growth in this sector.

India, with its rich biodiversity, abundant floral resources, and a deep-rooted tradition of natural remedies and perfumery, possesses unparalleled potential to become a global leader in the aroma industry. The country has a strong foundation of scientific know-how in agriculture, biotechnology, and fragrance technology, which can be harnessed to cultivate the finest flowers and develop world-class perfumes and aroma-based products.

What India needs now is a paradigm shift—a move from viewing flowers merely as ornamental or romantic symbols to recognising them as valuable business assets with immense commercial prospects. The missing pieces are innovative, mobile GMP-level processing units that can ensure high quality and hygiene standards, and a robust, global aggregator brand that can elevate Indian floral products onto the international stage.

Imagine a future where Indian farmers, entrepreneurs, and artisans collaborate to create exquisite perfumes, essential oils, and aroma products that rival the best in the world. These products can not only fetch premium prices but also promote sustainable livelihoods, rural entrepreneurship, and export growth.

The success of brands like Patanjali, which revolutionised the nutraceuticals and cosmetics sectors by emphasising quality, trust, and mass appeal, exemplifies what is possible when vision, innovation, and execution come together. Now is the time for a similar revolution in the floral and aroma industry. We must see flowers as a strategic business opportunity—an industry where India can lead with innovation, scale, and global branding.

By investing in modern processing technology, establishing quality standards, and developing a compelling global brand, India can turn its floral wealth into a world-class aroma industry. It's not just a romantic dream—it's a strategic economic opportunity that can elevate India's stature in the global natural product arena, create millions of jobs, and showcase the country's natural bounty and scientific prowess to the world. The moment to rise is now—growing high-quality flowers and crafting world-class perfumes can be India's next great success story, blending nature's beauty with timeless elegance.

•••

7

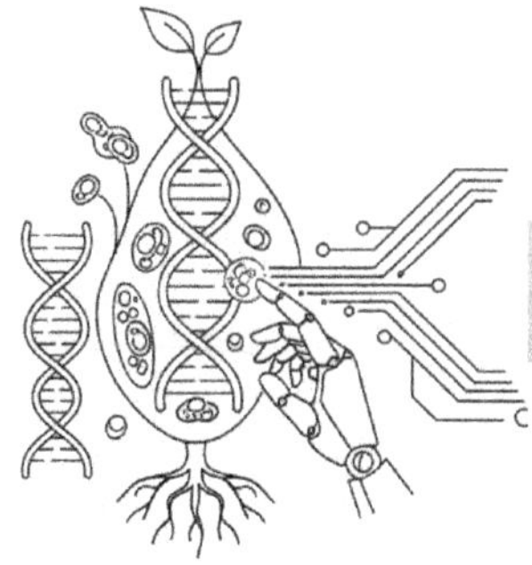

WONDER GRASS

In the hollow strength of bamboo lies a living metaphor — flexible yet unbreakable, ancient yet inexhaustibly new; it is the grass that can build homes, bind soil, fuel energy, and script the prosperity of villages reborn.

Bamboo has been a part of human habitat and livelihoods since ancient times, but has emerged as a remarkable resource with the potential to revolutionise the future of sustainable development and the global economy. Characterised by its rapid growth, versatility, and environmentally friendly attributes, bamboo offers a promising solution to many of the pressing environmental and economic challenges faced by the world today. Some people are even calling it the 'wonder grass'.

As if matching the fast-paced modern life, bamboo has an incredible growth rate. Unlike traditional timber, which can take decades to mature, bamboo can reach its full height within just 3 to 5 years. The full height of bamboo varies significantly depending on the species. *Dendrocalamus strictus*, also known as solid bamboo or male bamboo, typically grows up to 25 meters (about 80 feet). It may yield around 150-200 tonnes per hectare per year. *Bambusa vulgaris* Schrad. ex J.C. Wendl. is also common, reaching heights of over 20 meters, producing around 100-150 tonnes per hectare.

This rapid renewal cycle makes it an ideal renewable resource for bioenergy, paper, and other bio-based products, providing a continuous supply of raw material without depleting natural ecosystems. Bamboo can grow in diverse climates and poor soils and can thrive in areas unsuitable for other types of agriculture, making it an excellent sustainable resource. When harvested responsibly, bamboo can regenerate naturally without the need for replanting, thus reducing the ecological footprint associated with conventional wood harvesting.

Environmentally, bamboo plays a crucial role in combating deforestation, soil erosion, and carbon emissions. Its dense root system stabilises soil, preventing erosion, while its high photosynthetic efficiency enables it to absorb significant amounts of carbon dioxide from the atmosphere. Furthermore, bamboo is claimed to produce about 35% more oxygen than trees of the same size, contributing to improved air quality and climate regulation. A mature bamboo clump can produce approximately 20-30 tonnes of oxygen per hectare per year. Most of these claims are, however, anecdotal and need rigorous scientific validation in local conditions and specific varieties.

Economically, bamboo's versatility opens up a wide array of applications, making it a provider of livelihoods in rural economies. It can be transformed into a range of products, including furniture, flooring, paper, textiles, and even bioenergy. These industries not only promote local employment and income generation but also foster innovation in eco-friendly manufacturing processes. As global demand for sustainable products grows, bamboo-based industries are poised to become significant contributors to economic growth, especially in developing countries where resource optimisation is essential. Supporting bamboo cultivation and processing can promote rural development, reduce poverty, and enhance food security via intercropping and agroforestry systems.

During my January 2016 Taiwan visit, I saw the real potential of bamboo. When combined with new technology and creative ingenuity, it offered wonderful opportunities for wealth creation. The way it is used in traditional crafts, construction, furniture, paper, and art, symbolising sustainability and cultural heritage, was an eye-opener. Taiwan cultivates Moso bamboo (*Phyllostachys edulis* (Carrière) J. Houz.), also called Mao Zhu, through sustainable practices, promoting rapid growth, soil conservation, and biodiversity. It typically reaches 10 to 14 meters (30 to 50 feet), and culms (stems) can be 8 to 15 cm (3 to 6 inches) in diameter. New shoots emerging in spring can grow up to 1 meter per day!

The country's advanced processing techniques produce eco-friendly products, while bamboo's fast growth and biomass make it vital for reforestation, soil stabilisation, and carbon sequestration. Taiwan's innovations in green architecture are stunning. Bamboo-based materials are integrated into modern building designs for eco-friendly and earthquake-resistant structures. Taiwan is using bamboo for bioremediation to absorb pollutants from soil and water, contributing to environmental cleanup efforts. Furthermore, smart farming practices integrating IoT technology are being adopted to monitor and optimise bamboo growth, water usage, and pest control, thus increasing efficiency and reducing resource waste. Bamboos are indeed cared for and nurtured in Taiwan.

India is one of the largest producers of bamboo in the world, with bamboo growing naturally across states such as Northeast India, Odisha, Jharkhand, Madhya Pradesh, Karnataka, and Andhra Pradesh. The cultivation process involves selecting appropriate species, preparing the land, and planting rhizomes or seedlings, often utilising native bamboo varieties that are well-suited to local environmental conditions. But unlike in Taiwan, bamboos were taken for granted as something freely available. Only recently has the attitude changed.

The National Bamboo Mission (NBM) was first launched in 2006-2007, and the restructured NBM was launched by the Government of India in 2018-19 to promote bamboo cultivation, processing, and value addition. The mission aims to harness bamboo's potential as a sustainable and fast-growing resource to boost rural livelihoods, support agroforestry systems, and promote eco-friendly products. Besides increasing the bamboo cultivation area and improving the quality of bamboo, a robust market infrastructure for bamboo-based products is also being looked into. It encourages the establishment of bamboo-based industries, promotes natural bamboo regeneration, and supports the development of quality planting material.

Agronomically, bamboo belongs to the grass family (*Poaceae*). These plants typically have a perennial lifecycle, meaning they grow back year after year from the underground rhizomes, which spread horizontally to produce new shoots. The culms emerge from these rhizomes and can grow rapidly, often reaching their full height within a few months, depending on the species and growing conditions. Bamboo culms are hollow, segmented, and jointed, with nodes that provide structural support. The leaves are usually long, narrow, and arranged alternately along the culms, playing a role in photosynthesis.

Bamboo reproduces through two primary methods: sexual and asexual reproduction. Sexual reproduction occurs when bamboo species flower, a process that can take between 3 to 120 years depending on the species. When flowering, many bamboo plants in a large area often flower simultaneously, a phenomenon known as monocarpic flowering. Following this, bamboo produces seeds that require specific conditions, such as moisture and favourable temperatures, to germinate and grow into new plants. This flowering cycle is typically synchronised across vast regions, leading to mass seeding events followed by the dieback of the flowering culms.

Bamboo also reproduces asexually through underground rhizomes, horizontal stem-like structures that spread beneath the soil. New culms emerge from these rhizomes, forming dense groves and enabling rapid clonal propagation. This method of vegetative growth allows bamboo to quickly colonise available spaces and establish extensive stands, often resulting in large, uniform bamboo forests. Generally, reproduction through rhizomes is the primary mode for most bamboo species, accounting for approximately 80-90% of their propagation. The seed production typically results in the death of a significant portion of the mature culms, after which new growth primarily relies on rhizome propagation. The combination of infrequent but synchronised flowering and efficient vegetative spread ensures bamboo's resilience, longevity, and ability to adapt to various environments. Generally, bamboo can emerge from the ground within 3 to 6 weeks after planting or harvesting. The new culms typically reach their full height within 3 to 4 months for most fast-growing species, although some varieties may take longer, up to 1 year or more, to fully mature.

Edible bamboo mainly refers to tender bamboo shoots. Rich in fibre, low in calories, and containing essential nutrients such as vitamins and minerals, these are widely consumed in various Asian cuisines, including Chinese, Japanese, Korean, Thai, and Indian foods. Bamboo shoots contain globulins and albumins, similar to those in legumes and vegetables. These proteins are easily digestible, making bamboo a good source of vegetarian protein. Besides their nutritional value, bamboo shoots add a crunchy texture and mild flavour to dishes like stir-fries, soups, pickles, and salads.

Bamboo leaves and extracts have been studied for their medicinal value. Notably, they contain flavonoids such as orientin, vitexin, and isoorientin, which exhibit strong antioxidant effects that help neutralise harmful free radicals in the body. Additionally, phenolic acids like gallic acid, ferulic acid, and chlorogenic acid contribute to their anti-inflammatory and antimicrobial activities. The broad class of polyphenols present in bamboo further enhances these health benefits by supporting immune responses and reducing inflammation. Moreover, polysaccharides found in bamboo extracts may boost immune function, while catechins such as epicatechin are recognised for their potent antioxidant activity. Collectively, these compounds make bamboo leaves a promising natural source for combating oxidative stress, infections, and inflammatory processes, with potential applications in health and medicine.

Bamboo charcoal is widely recognised for its detoxification properties and benefits for oral health. Its porous structure allows it to effectively adsorb toxins, impurities, and unwanted substances from the body, making it popular in detox products. In oral health, bamboo charcoal is commonly used in toothpaste and mouth rinses due to its ability to gently absorb bacteria and freshen breath. Its natural antibacterial and deodorising properties help reduce plaque buildup and promote healthier gums. Additionally, activated bamboo charcoal removes stains from tooth enamel.

Bamboo extract has been studied for its potential benefits in managing diabetes, primarily due to its rich content of antioxidants, polyphenols, and various bioactive compounds. These substances may help improve insulin sensitivity, reduce oxidative stress, and lower blood sugar levels. Some research suggests that bamboo extract can inhibit enzymes that break down carbohydrates, thereby slowing glucose absorption and helping regulate blood glucose levels more effectively. Additionally, its anti-inflammatory properties may contribute to reducing chronic inflammation associated with insulin resistance. While more clinical studies are needed to fully establish its efficacy, bamboo extract shows promise as a natural supplement to support diabetes management and overall metabolic health.

The high cellulose content in bamboo, typically 40% to 50% of its dry weight, makes it an ideal raw material for producing bioplastics, which are biodegradable and sustainable alternatives to conventional plastics derived from fossil fuels. The process involves breaking down bamboo's lignocellulosic biomass into fermentable sugars, primarily glucose, through a series of steps including pretreatment, enzymatic hydrolysis, and fermentation. During fermentation, specific microorganisms convert the glucose into lactic acid. This lactic acid is then purified and polymerised to produce polylactic acid (PLA), a biodegradable polyester made up of repeating lactic acid monomer units, $(C_3H_4O_2)n$, with n as the degree of polymerisation.

Integrating bamboo with three sister crops (*tres hermanas* in Spanish) of the native people of Central and North America, namely squash (*Cucurbita maxima* Duchesne), maize (called corn), and climbing beans [typically tepary beans (*Phaseolus acutifolius* A. Gray) or common beans (*Phaseolus vulgaris* L.)] promotes complementary growth. Native American groups also planted sunflowers (*Helianthus annuus* L.) on the north edges of their gardens as a "fourth sister". This "return to old school" would not only enhance agroforestry systems by promoting biodiversity, reducing soil erosion, and optimising land productivity, but also maximise economic benefits and reduce dependency on a single resource. Other companion crops include millets [general mention; *Pennisetum glaucum* (L.) R.Br. (pearl millet, example)], which can be grown around bamboo [general mention provided later; *Bambusoideae* (*Poaceae*, multiple genera)] groves without competing for resources, providing additional income for farmers. The imagery of the bamboo as the protective brother of four sisters providing a full course of nourishing food is indeed inspiring.

The biomass grass economy is seen as a vital part of the transition to a circular, sustainable, and low-carbon economy. It offers rural development opportunities while supporting clean energy initiatives globally. Miscanthus (*Miscanthus* × *giganteus* J.M. Greef & Deuter ex Hodkinson & Renvoize, elephant grass), though not yet popular in India, is known for its high biomass yield and is used for bioenergy production, biofuels, and bioproducts. It can grow on marginal lands, making it an environmentally sustainable bioresource. Kikuyu grass (*Pennisetum clandestinum* Hochst. ex Chiov.) is a fast-growing perennial grass used as a nutritive fodder. It is also used for preventing soil erosion due to its deep (up to 3 m), strong root system. It is also a candidate for phytoremediation of saline soils and has some minor traditional medicinal uses. The plant originated from the highland pasture land of East Africa, and it tolerates low temperatures. The plant originated in the highlands of East Africa and was introduced as a fodder crop in several countries, including India. Since it grows rapidly in challenging conditions, it is also a candidate for biomass production for energy.

Vetiver [*Chrysopogon zizanioides* (L.) Roberty (syn. *Vetiveria zizanioides* (L.) Nash)] is valued for its strong root system that is used in soil erosion control, phytoremediation, and as a raw material in aromatic oils and perfumes. Its biomass also has potential for bioenergy applications and is predominantly grown in Kerala, Tamil Nadu, Andhra Pradesh, Odisha, West Bengal, Uttar Pradesh, Maharashtra, and Gujarat. Switchgrass (*Panicum virgatum* L.) is a native North American grass and has not yet been widely cultivated or naturally established in India. However, there is growing interest in exploring its potential as a biomass crop in India due to its high yield, adaptability, and suitability for bioenergy production.

Bamboo is increasingly recognised as a vital sustainable construction material in the low-carbon era due to its numerous environmental, economic, and structural benefits. One of the most significant advantages of bamboo is its rapid growth; it can grow up to one meter per day, making it a highly renewable resource that can be harvested sustainably without depleting natural ecosystems. Additionally, bamboo has a high capacity for carbon sequestration, absorbing more CO_2 than many other plant species, which helps mitigate climate change. Its minimal requirements for fertilisers and pesticides further contribute to its low environmental impact, making it an eco-friendly alternative to conventional building materials.

In terms of structural properties, bamboo is known for its high strength-to-weight ratio, providing durability and flexibility suitable for various construction applications. It is lightweight yet strong enough to be used in load-bearing walls, scaffolding, flooring, roofing, and even entire structures. The cost-effectiveness of bamboo also adds to its appeal, especially in developing regions where affordability is crucial. Moreover, bamboo is biodegradable, which means it reduces pollution when used as a building material. Its resilience to natural disasters, such as earthquakes and storms, makes it an excellent choice for disaster-resilient construction, promoting safer and more sustainable urban environments.

In a low-carbon future, bamboo plays a vital role by supporting green architecture and energy-efficient building practices. Its use in eco-friendly construction reduces the embodied energy associated with materials like steel and concrete. Bamboo structures aid in lowering overall carbon emissions and contribute to climate change mitigation efforts. However, to maximise its benefits, proper plantation management and harvesting practices are essential to ensure sustainability.

As building codes and standards evolve, integrating bamboo into mainstream construction will become more feasible, especially with advances in treatment and preservation techniques that enhance its durability in various environments. Overall, bamboo stands out as a promising, sustainable material that can significantly contribute to environmentally responsible development in the low-carbon era. Rajveer Singh Brar, a farmer in Fazilka, Punjab, testifies from his experience as we quote below:

> Environmentally, it's a powerhouse —absorbing 4–5 times more carbon dioxide and releasing 35% more oxygen than most plants, while drawing moisture from the atmosphere and storing it in the soil and one can get around 100 tonnes of bamboo per hectare per year without replanting it for a century as against 25 to 30 tonnes per hectare for other trees that need re-transplantation every 5-10 years. The initial cost is Rs 40,000–50,000 per acre, after which maintenance is minimal. "No fertilisers, no pesticides—bamboo practically grows on its own after establishment."[1]

1 https://indianexpress.com/article/cities/chandigarh/at-the-centre-of-punjab-farmers-vrs-plan-a-fast-growing-grass-he-calls-green-gold-10194210/

IIT Kanpur has been involved in the development of bamboo-based crockery as a sustainable alternative to traditional materials like plastic and ceramic. Bamboo crockery is an eco-friendly alternative to traditional plastic and ceramic tableware, made from processed bamboo fibres and often blended with other plant-based materials. This innovative material is lightweight, durable, and can withstand high temperatures, making it microwave and dishwasher-safe. It's also a sustainable choice, as bamboo is a renewable resource that grows quickly and requires minimal resources to cultivate.

Vegan leather made from biomass is an innovative and sustainable alternative to traditional animal leather. It is typically produced using plant-based materials or renewable biomass sources such as mushroom mycelium, cork bark, pineapple leaf fibres, or cactus (*Opuntia ficus-indica* (L.) Mill.) plants. Mushroom (usually *Ganoderma lucidum* (Curtis) P. Karst. or *Pleurotus ostreatus* (Jacq.) P. Kumm.) leather, created from the root structure of fungi, is biodegradable, durable, and grows quickly in controlled environments. Cork leather is derived from the bark of cork oak trees, making it a renewable and lightweight material with a leather-like appearance. Pineapple leather, known as Pinatex, is produced from the fibres of pineapple leaves, which are a byproduct of agriculture and help reduce waste. Cactus leather is made from cactus plants that require minimal water and no pesticides, making it environmentally friendly. Additionally, some vegan leathers are created from bio-based plastics like bio-PVC or bio-TPU, derived from biomass sources, offering synthetic but eco-friendly options. These biomass-based vegan leathers are gaining popularity among brands committed to sustainability and ethical practices, as they often replicate the look and feel of traditional leather while significantly reducing environmental impact and avoiding animal suffering.

In Europe, countries like Italy, the Netherlands, and Germany have made significant advances with startups and established companies developing plant-based and bio-based leathers, often driven by stringent sustainability and environmental regulations. The Netherlands hosts several innovative firms working on mushroom leather and other bio-materials, while Italy, known for its leather craftsmanship, is increasingly adopting vegan and sustainable alternatives. Germany is also leading in synthetic and bio-based leather innovations, supported by strong technological research and manufacturing capabilities. Overall, these regions are at the forefront of mature vegan leather technology, with ongoing advancements and growing adoption across industries.

Gujarat, which is a hub for leather and textile industries, as well as Bangalore and Delhi, where innovative startups and research centres are working on sustainable biomaterials, can adopt vegan leather. Companies like Pinatex India are exploring pineapple leaf fibres, while other startups focus on mushroom leather and cactus-based products. Additionally, institutions like the Indian Institute of Technology (IITs) and various universities are conducting research on bio-based materials and sustainable bioproducts. Overall, India is gradually advancing in this field, driven by the growing demand for eco-friendly and ethical alternatives to conventional leather.

A futuristic grass-based building material is likely to be a sustainable, eco-friendly, and innovative composite derived from various grass species such as bamboo, vetiver, or switchgrass. These materials could include advanced bio-composites made by combining grass fibres with biodegradable resins or polymers to create lightweight, strong, and durable panels or tiles. Such materials might feature self-healing properties, insulation capabilities, or even integrated renewable energy systems like photovoltaic fibres. Additionally, 3D-printed structures using grass-based bio-inks could allow for customisable, biodegradable, and environmentally sustainable architecture. These futuristic materials would prioritise carbon sequestration, low water usage, and minimal environmental impact, making them ideal for the next generation of eco-friendly construction.

But while there are immense possibilities, there is also a heavy dependency on imports for certain inputs. Take the example of the core or the bamboo sticks used as the base for incense sticks, which are often imported into India largely from Vietnam and China. India primarily produces the aromatic ingredients and the mixture of herbal and natural compounds used in incense sticks domestically, but the bamboo sticks serve as the support structure and are frequently sourced from abroad due to cost, quality, or availability factors. Some Indian manufacturers also produce bamboo sticks locally, but a significant portion of imported bamboo is used in the incense stick industry.

I consider the Northeast region, including states like Assam, Arunachal Pradesh, Nagaland, Manipur, Mizoram, and Tripura, to be ushering in a "Bamboo Revolution". These states have the highest bamboo diversity in the country, with numerous species naturally occurring in forested and hilly terrains. The region's favourable climate and dense forests provide ideal conditions for bamboo growth, making it a vital resource for local livelihoods, construction, handicrafts, and ecological stability. The key would be to bring in technology to use the bamboo grown here and convert it into high-value products.

High-value materials derived from bamboo include a range of innovative and premium products that capitalise on bamboo's strength, flexibility, and sustainability. Some notable examples are bamboo-based bio-composites used in automobile parts, furniture, and flooring, which combine bamboo fibres with resins to create lightweight, durable, and eco-friendly panels. Bamboo charcoal and activated bamboo charcoal are valued for their uses in air and water purification, as well as in health and beauty products. Bamboo fibre textiles, which are soft, breathable, and strong, are considered luxury fabrics in fashion and interior design. Additionally, bamboo-based bio-manufactured materials such as biodegradable plastics, paper, and packaging materials are gaining high value due to their environmental benefits. Lastly, bamboo-derived nanomaterials and carbon nanotubes are emerging as high-tech, high-performance materials used in electronics, aerospace, and advanced composites.

Vegan leather shoes will become a new fashion and can be made here and sold globally.

I see events like the national conference on dissemination of innovations in the bamboo sector for improving the rural economy, held in March 2022 at the National Institute of Rural Development and Panchayati Raj, Hyderabad, as an auspicious beginning of a new era. The bamboo market worldwide is about $75 billion, projected to reach $100 billion by 2030. China has a 70% share of this, whereas India contributes only 5%. A determined and integrated approach of infusing biotechnology, financing, and economic translation into industry can capture a quarter of the global market in the next five years.

The innovative use of bamboo in a bio-conscious society offers a compelling pathway towards sustainable development and socio-economic upliftment. Bamboo is a versatile, fast-growing, and environmentally friendly resource. Its ability to grow rapidly—reaching maturity within just three to five years—makes it an ideal renewable substitute for traditional timber, plastics, and other non-biodegradable materials. Cultivating bamboo requires minimal chemical inputs and helps prevent soil erosion, contributing positively to ecological stability. By harnessing bamboo's potential, societies can reduce reliance on environmentally harmful resources while supporting a circular economy centred around sustainability.

This approach can significantly impact rural livelihoods by creating new economic opportunities. Local communities can establish industries focused on bamboo products such as furniture, construction materials, handicrafts, paper, and textiles. Such ventures not only generate income but also offer skills development, fostering entrepreneurship among rural populations. This economic diversification provides a powerful incentive for young people to remain in their villages rather than migrating to overcrowded cities, where urban challenges like crime and congestion tend to rise. Through these initiatives, rural areas can experience a renaissance, transforming from neglected regions into thriving hubs of eco-friendly enterprise and innovation.

Additionally, promoting bamboo cultivation and processing has notable environmental benefits. Bamboo acts as a natural carbon sink, absorbing substantial amounts of CO_2 and helping combat climate change. Its extensive root system stabilises the soil, preventing erosion and desertification, which further preserves and restores local ecosystems. The environmental advantages of bamboo align seamlessly with the goals of a bio-conscious society, emphasising harmony between development and nature. Moreover, bamboo's applications extend into urban environments through sustainable architecture, eco-friendly furniture, and green infrastructure projects that reduce reliance on high-carbon-footprint materials like concrete and steel.

Addressing societal challenges such as overcrowding and urban crime, bamboo-based rural industries can serve as a catalyst for social stability. Harnessing bamboo's multifaceted potential not only fosters environmental sustainability but also drives economic resilience and social transformation, making it a visionary strategy for building healthier, more sustainable societies. I close this chapter by suggesting a five-dimensional initiative for the inspired entrepreneurs: Activated carbon (AC), bamboo fibres, disposables & pencil body, food-grade silica from bamboo leaves, and fire-retardant and anti-bacterial coatings for bamboo structures. These are not mere ideas but scientifically established possibilities.

The iodine number is a key parameter used to evaluate the adsorptive capacity of activated carbon, specifically its surface area and porosity. According to American Society for Testing and Materials (ASTM) standards, activated carbon with an iodine number between 600 and 1,450 mg/g is considered of high quality and suitable for various filtration and adsorption applications. For example, activated carbon with an iodine number above 600 mg/g is generally regarded as having good adsorptive properties, and those approaching 1,450 mg/g are considered high-capacity, suitable for specialised applications like gas purification and water treatment (ASTM D5832).

The most well-known applications of activated carbon are water and air purification, as well as the food, pharmaceutical, and electronics industries. Activated carbon is obtained from the controlled pyrolysis of a raw material (bamboo), where temperatures vary from 400 to 1,200 °C. It has been synthesised from about 35 different species of bamboo, and its various properties, namely iodine number, calorific value, and ash content, have been established. The carbonisation capabilities have been optimised according to the age and moisture content of bamboo. Carbonisation furnaces with a processing capacity of about 1 tonne are now commonplace.

Generally, fibre extraction of bamboo is done through mechanical (enzymatic) or chemical methods. It is also done through a new technique of sonication, which uses high-frequency sound waves, typically in the ultrasonic range, to induce physical and chemical changes in materials. When bamboo is immersed in water, the ultrasonic waves create rapid pressure fluctuations and cavitation bubbles in the liquid. As these bubbles collapse, they generate localised high temperatures and intense shear forces. These physical effects help break down the complex matrix of bamboo at a microscopic level, facilitating the separation of fibres from the woody tissue, significantly enhancing fibre extraction by loosening the natural bonds and making it easier to obtain clean, long, and uniform bamboo fibres.

Agarbatti stick manufacturing and artisan-making industries generate about 60-80% of bamboo dust as waste. This dust is effectively utilised to convert it into value-added products like pencil bodies and disposable cups and plates, which are completely biodegradable in nature. Bamboo leaves are a very good source of silicon dioxide, or silica, and they have a wide utility, from medical to agricultural applications. Bamboo structures and bamboo-based furniture are easily attacked by mould and bacteria, which reduces their lifespan. Moreover, bamboo structures are easily prone to fire. Silica-based fire-retardant coatings provide thermal stability and create a barrier that resists flames and heat spread.

Aluminium-based coatings, which are reflective, can also offer some degree of fire resistance by reflecting heat away from the surface. Chemical treatments such as phosphorus-based and boron-based fire retardants are often incorporated into varnishes or paints; these promote char formation and inhibit flame spread by limiting oxygen access. Additionally, intumescent varnishes are specially designed to swell upon heating, forming an insulating layer that protects the bamboo. It is important to choose coatings that are compatible with bamboo and adhere to safety and environmental standards, often requiring consultation with manufacturers or fire safety experts to select the most effective and eco-friendly options.

Like Taiwan revolutionised wealth through chip manufacturing, Northeast India has the potential to harness bamboo as a transformative, green resource. By embracing modern technology and innovative applications, the region can elevate bamboo from traditional uses to high-value, sustainable products such as bio-composites, eco-friendly construction materials, bioplastics, and vegan leather. With its vast bamboo diversity, favourable climate, and rich natural resources, Northeast India can lead in developing cutting-edge bamboo-based industries that generate economic growth, promote environmental conservation, and create numerous livelihoods. This shift can position the region as a global hub for sustainable materials, aligning economic progress with ecological preservation.

•••

Gradually, a collaborative community get built, both with other educators and with our students, to foster learning and support within the laboratory and field environment.

In the presence of our esteemed mentor, Dr Lok Man Singh Palni, we celebrate the timeless spirit of our institution—nurturing what he once planted, and watching it flourish.

Veteran geneticist and former Director-General of the Indian Council of Agricultural Research, Dr Virender Lal Chopra, and the narrator of this book, Prof. Arun Tiwari, came to CSIR Foundation Day in 2018.

My guide during my brief stay at JNU, Dr Sudhir Kumar Sopory became my torchbearer. A fellow of The World Academy of Sciences (TWAS), he later became the Vice Chancellor of Jawaharlal Nehru University.

Dr Girish Sahni, Director-General, CSIR, led me like a colleague and supported CSIR-IHBT to help achieve its fullest potential. Seen here are Dr Shyam Kumar Sharma (on the right) and Dr Minto Purshotam Gupta (on the left).

Dr Shekhar C. Mande, Director-General, CSIR, led us from the front. Not only was he always there to guide and support, but he also helped expand our vision.

The support of Dr Nallathamby Kalaiselvi, Director-General, CSIR, reflects the vital role of top leadership in sustaining ground-level initiatives and scaling our work into impactful national efforts

Growing apples in Manipur was a crown jewel of our work.
Where apples were not even reachable by road,
they were grown in plenty, creating a new source of income for farmers.

We grew the first monk fruit in India. Renowned for its anti-inflammatory properties, monk fruit derives its sweetness from mogrosides and is considered safe for individuals with diabetes.

Heeng was successfully grown in the Lahaul Valley. India spends $100 million every year to import 1,500 tons. In future, it will remain with our farmers.

We introduced true cinnamon cultivation in five districts of Himachal Pradesh —Una, Hamirpur, Bilaspur, Kangra, and Sirmaur—to boost domestic production and reduce India's reliance on imports.

Indian Tulips at CSIR-IHBT is the shining example of selecting suitable agro-climatic conditions coupled with hydroponics for off-season cultivation. We could stop importing tulip bulbs to keep our gardens blooming.

Two distinguished leaders, Shri Shivraj Singh Chouhan (then Chief Minister of Madhya Pradesh) and Dr Jitendra Singh, Union Minister, at the CSIR pavilion at Bhopal, celebrate the Purple Revolution — where lavender blooms brought prosperity to farmers, turning fragrance into the true scent of success.

Your work becomes your stature. With stalwarts of Indian science, Dr Ashwani Pareek, Dr Cherukumalli Srinivas Rao, Dr Soumya Swaminathan, Dr Renu Swarup, Dr Trilochan Mohapatra, (and to my right) Dr Kailash Chander Bansal, and Dr George John.

With the doyens of Indian agricultural research. Legendary Dr Rajendra Singh Paroda and the Secretary of the Department of Agricultural Research and Education (DARE) and Director-General, ICAR, Dr Mangi Lal Jat. I have merely lived a legacy.

You don't choose your family. They are God's gift to you, as you are to them. Like planets in space hold each other, every individual's work is possible only with a supporting family.

8

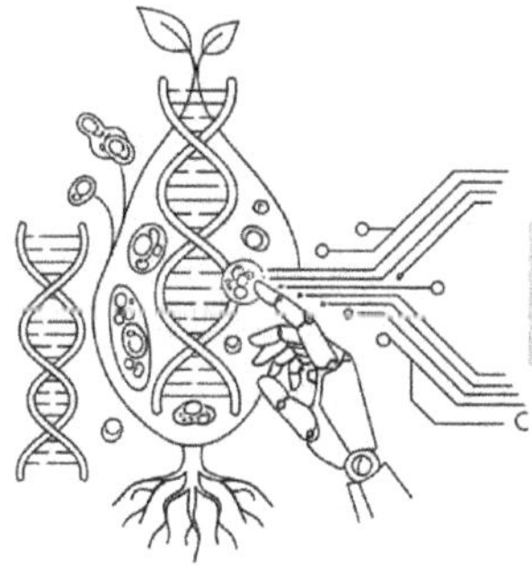

VALUE-ADDITION

Value is born not in what the earth gives alone, but in how wisely we transform and share it.

East Asian agriculture exemplifies a harmonious balance between honouring the past and embracing the future. This chapter delves into how these combined efforts are transforming the landscape of food production, ensuring both cultural preservation and economic resilience. Farmers and researchers blend traditional knowledge with cutting-edge biotechnology, sustainable practices, and digital technology. This approach respects the cultural roots and sustainable wisdom embedded in traditional methods while integrating new technologies to boost productivity, resilience, and environmental stewardship. By bridging the gap between the past and the present, farmers and researchers are redefining what it means to honour tradition in a rapidly changing world, creating a compelling narrative of a modern bioeconomy rooted in history.

Earlier, Thailand was known as Siam. It was predominantly an agricultural society along the Chao Phraya and the Mekong Rivers. Thailand's rice farming was more influenced by the Chao Phraya River plains. These early inhabitants developed sophisticated techniques for rice cultivation, including irrigation systems, enabling them to sustain growing populations and establish stable communities. Rice cultivation became the foundation of Thailand's economy and way of life. It was not only the staple food but also a vital commodity for trade and cultural practices. During the reign of the Sukhothai and Ayutthaya kingdoms (from the 13th to the 18th centuries), agriculture flourished under a feudal system where land was worked by farmers under royal authority, supporting the prosperity of these kingdoms through local and regional trade.

As Thailand moved into the 19th and 20th centuries, efforts to modernise agriculture gained momentum. The introduction of new crop varieties, farming tools, and irrigation projects helped increase productivity. The government initiated land reforms and infrastructure improvements to support rice production, positioning Thailand as a major rice exporter in the global market. Over time, other agricultural products such as rubber, tapioca, sugarcane, and various fruits also became important, further bolstering the nation's economy. Modern policies now focus on promoting sustainable farming practices and diversifying income sources for farmers, aiming to balance economic growth with environmental preservation.

Cassava plays a vital role in Thailand's bioeconomy, serving as a key raw material for various industries such as bioenergy, food, and manufacturing. The country harnesses cassava primarily for bioethanol production, contributing to renewable energy initiatives, and for processing into tapioca starch, which is widely used in food products like bubble tea, desserts, and snacks. Additionally, derivatives from cassava are employed in creating biodegradable plastics, adhesives, and paper, further expanding its industrial applications. The cultivation and processing of cassava generate substantial income for farmers and support rural development, making Thailand one of the world's leading exporters of cassava and its products. However, the industry faces sustainability challenges, including land use concerns, environmental impacts, and market price fluctuations. As a result, ongoing efforts aim to promote sustainable cultivation practices and ensure the long-term contribution of cassava to Thailand's bioeconomy.

Thailand is a great model of sustainable farming practices, methods of agriculture that aim to meet current food and fibre needs without compromising the ability of future generations to do the same. Techniques such as organic farming, crop rotation, integrated pest management, water conservation, and soil preservation are a way of life in Thailand. These practices help improve soil fertility, reduce pollution, and ensure that water resources are used efficiently, ultimately supporting the long-term productivity of agricultural land.

Diversifying income sources for farmers is another strategy being promoted in Thailand to enhance economic resilience and rural development. Instead of relying solely on traditional crops like rice, farmers are encouraged to cultivate a variety of crops, engage in livestock farming, or develop agro-tourism and value-added products. This diversification helps reduce risks associated with market fluctuations, climate change, and crop failures. By diversifying their income, farmers can stabilise their livelihood, increase household income, and build resilience against economic and environmental challenges.

Thailand is one of the world's leading producers and exporters of natural rubber, and the crop has played a significant role in its agricultural economy for many decades. The French colonisers, who controlled parts of Indo-China (including present-day Vietnam, Cambodia, and Laos), were among the first to recognise the potential of rubber as a lucrative crop. Rubber cultivation in Thailand began in the late 19th century, initially introduced from British Malaya by Phraya Ratsadanupradit as a way to diversify agriculture and tap into the growing global demand for rubber, particularly in the automotive and tyre industries.

The expansion of rubber plantations took off after World War II, driven by increasing international demand. Thailand's tropical climate and suitable soil conditions provided ideal environments for rubber tree cultivation. Most rubber plantations are concentrated in the southern part. Instead of just exporting raw rubber, Thailand has developed advanced rubber processing industries, producing various rubber products, including sheets, latex, and tyres, for the global market.

The growing global demand for rubber, driven by the automotive and tyre industries, encouraged further expansion of rubber plantations in the region. The development of rubber estates provided employment opportunities but also brought challenges such as land displacement and environmental concerns. Neighbouring Vietnam became a significant rubber producer. Over time, technological advancements improved rubber production and processing, making Indo-China a significant player in the global rubber market. In the post-colonial era, Vietnam became one of Asia's leading rubber producers with plantations operated by both state and private entities.

Innovation is not merely about science or methods. It also means opening up to new areas. The way Vietnam turned into one of the world's leading producers and exporters of cashew nuts is a great example of innovating tradition. Vietnam's favourable climate, suitable soil conditions, and government support helped establish extensive cashew plantations, particularly in the Central and South Central regions. The country focused on improving cultivation techniques, expanding processing capacity, and establishing a robust export market. As a result, Vietnam tops as a cashew exporter, while the Ivory Coast (officially Côte d'Ivoire, a country in West Africa) is the world's largest raw cashew producer. India plays a major role in cashew export, but depends on the import of large quantities of raw cashews from the Ivory Coast and Tanzania to feed the processing industry.

Vietnam primarily imports raw cashew nuts in large quantities to meet its processing needs. The country is one of the world's leading processors and exporters of cashew products, transforming raw nuts into roasted, salted, and packaged cashews for global markets. By importing raw cashews from countries like the Ivory Coast and Nigeria, Vietnam leverages its advanced processing industry to add value, generate export revenue, and support economic growth. This import-and-process model, a shining example of value-addition, has established Vietnam as a major player in the global cashew supply chain.

As one of the leading global exporters of processed cashew nuts, Vietnam has experienced substantial growth in rural incomes and employment opportunities, particularly in regions dedicated to cashew cultivation and processing. The development of this industry has not only increased foreign exchange earnings but also fostered the growth of related sectors such as logistics, packaging, and trade. Overall, cashew processing has become a vital driver of economic prosperity, contributing to improved living standards and sustainable development in Vietnam.

India and Indonesia share a deep historical oneness rooted in centuries of cultural, religious, and trade exchanges. Maritime trade routes facilitated the flow of ideas, goods, and explorers between these regions, fostering a sense of interconnectedness. Additionally, both countries were influenced by Indian civilisation, which played a key role in shaping their cultural identities, language, and agricultural practices. Airlangga, which means "The Preserver", was a very famous Hindu king ruling over the region in the 11th century. One can see ancient temples that date back to the early common era in every major island of the archipelago.

In the post-World War II era, Indonesia has emerged as the world's largest producer of commodities like palm oil, rubber, spices, and coffee, and invested in value addition. The country has focused on developing industries such as food processing, herbal and medicinal products, and packaged agricultural goods. For example, instead of exporting raw cocoa beans, Indonesia now produces processed chocolate and cocoa-based products, which fetch higher prices on the international market.

The government has promoted value-added agriculture through policies and investments in infrastructure, technology, and training. This includes establishing processing plants, improving supply chains, and encouraging entrepreneurship in agricultural sectors. In particular, industries like palm oil refining, coffee roasting, and spice processing have seen significant growth. It provided multiple benefits, including increased local employment, improved farmers' incomes, and greater competitiveness in global markets.

Makhana and Sabudana are perhaps the best examples of value addition in India. Also known as fox nuts, makhana is a highly valued traditional agricultural product with significant nutritional and medicinal benefits. It is derived from the seeds of the water lily plant (*Euryale ferox* Salisb.), which grow in freshwater ponds and wetlands. Makhana is widely cultivated in regions such as Bihar, Jharkhand, Uttar Pradesh, and parts of Assam and Manipur in India. Makhana seeds are rich in protein, fibre, antioxidants, and minerals, making them a healthy option for weight management, heart health, and digestive health.

The cultivation of makhana is eco-friendly and promotes water conservation, as it is grown in aquatic environments that support freshwater ecosystems. The process involves pond-based farming, which can improve the livelihood of local farmers by providing an alternative source of income, especially during off-season periods for other crops. Recently, there has been a push to promote makhana as a value-added product through processing, packaging, and export. It is now available in various forms such as roasted snacks, flour, and health supplements. This has helped create a niche market for makhana globally, boosting rural economies and encouraging sustainable farming practices. Overall, makhana holds great promise as a nutritious, eco-friendly crop with the potential to enhance rural income, promote health, and support sustainable water-based agriculture.

Sabudana, also known as tapioca pearls or sago, is a popular food ingredient derived from the starch of cassava (*Manihot esculenta* Crantz) root. It is widely used in Indian cuisine, especially during fasting periods and festivals, as it is considered a wholesome and easily digestible carbohydrate source. The production of sabudana involves extracting starch from cassava roots, which are processed into fine flour or starch. This starch is then soaked and shaped into small pearls or balls, which are dried and packaged for consumer use. India is a major producer and consumer of sabudana, with the main cultivation zones located in parts of Maharashtra, Gujarat, and the Northeastern states. While sago can sometimes refer to starch from sago palm (*Metroxylon sagu* Rottb.), in India, "sago" is used to describe cassava-derived pearls.

Sabudana is valued not only for its culinary uses but also for its nutritional properties. It provides energy due to its high carbohydrate content and is easy to prepare—commonly eaten as kheer, vada, or consumed directly after soaking, especially during fasting. Additionally, being gluten-free and easy to digest, it is suitable for dietary restrictions or those with gluten sensitivities. With increasing awareness of healthy eating and gluten-free diets, the demand for sabudana has risen both domestically and internationally. Efforts are underway to promote sustainable cassava cultivation to meet the growing demand while ensuring resource conservation and environmentally friendly farming practices. Overall, sabudana remains an important part of traditional cuisine and a significant agricultural product linked to the cassava farming industry.

The story of the chestnut (*Castanea* spp.) is, however, a little sordid because of the incidence of chestnut blight (a fungal disease) that destroyed billions of trees. Chestnuts, called *shahbaloot* are produced by trees belonging to the genus *Castanea*. Rounded or slightly elongated with a smooth, shiny shell, the nut is usually 2-4 cm in diameter, sweet, starchy, and delicately flavoured with a kernel inside the shell. Rich in carbohydrates, dietary fibre, vitamins such as vitamin B and C, and minerals like potassium and magnesium, chestnuts are popular worldwide. Chestnut trees are found in various regions with temperate climates worldwide. China, Korea, Japan, Italy, Turkey, France, and parts of North America grow chestnut. Commonly cultivated species include the European chestnut (*Castanea sativa* Mill.), American chestnut (*Castanea dentate* (Marshall) Borkh.), and Chinese chestnut (*Castanea mollissima* Blume). In India, varieties of both European and Chinese chestnut are found in the Himalayan regions, including

Himachal Pradesh, Jammu and Kashmir, Uttarakhand, and parts of Northeast India. While they may currently be less popular than cashews due to limited awareness and susceptibility to pests and diseases, with improved planting material, cultivation techniques, and proper marketing, I see chestnuts as a money spinner.

When I connect all these dots, a beautiful model of the 21st-century agribusiness model emerges, which emphasises sustainability, technology, diversification, and value addition to meet global demands. The secret lies in integrating innovative practices and digital tools to improve productivity, efficiency, and profitability for farmers, bringing their products to the global market as value-added products rather than raw materials. The prudent Europeans initially created regulatory barriers to agricultural imports so that they buy only the best and reject the rest.

But as technology proliferated, the adoption of precision agriculture, which uses GPS, sensors, and data analytics to optimise the use of water, fertilisers, and pesticides, met whatever regulations demanded and even more. As an unintended bargain, this not only increases crop yields but also minimises environmental impact. Additionally, the use of biotechnology and improved seed varieties helps enhance disease resistance and climate adaptability.

Another important feature is the integration of supply chains and market access through digital platforms, e-commerce, and smart logistics. This allows farmers to connect directly with consumers and global markets, reducing middlemen and increasing profit margins. Diversification of crops and livestock to reduce dependency on a single commodity also plays a critical role, enabling farmers to mitigate risks related to price fluctuations or climate change.

Sustainable practices, including organic farming and resource conservation, are central to the modern agribusiness model. The focus on value addition—processing raw agricultural products into packaged, branded goods—creates higher income streams and greater market competitiveness. I share here four shining examples of the 21st-century agribusiness model that showcase innovation, sustainability, and value addition. Each model can be replicated in India, offering unparalleled investment opportunities and lucrative profit margins.

1. Drip Irrigation in Israel – Israel's widespread adoption of precision irrigation technology has revolutionised water use efficiency in agriculture. This sustainable method increases crop yields while conserving water, exemplifying technological innovation in farming. Developed in the 1960s by Israeli researcher of Polish origin Simcha Blass (1897 –1982) and his son Yeshayahu, and the company Netafim at Kibbutz Hatzerim (established 1965) commercialised it. Drip irrigation involves delivering water directly to the root zone of plants through a network of hoses, emitters, and pipes. This targeted approach minimises water waste and ensures that crops receive a consistent and precise amount of moisture. This led to the establishment of an oasis called a Kibbutz, at Hatzerim, located in the Negev desert in Israel.

Israel, a country characterised by arid and semi-arid climate conditions with limited water resources, faced significant challenges in maintaining sustainable agriculture. The adoption of drip irrigation revolutionised farming in the region, enabling farmers to dramatically increase crop yields while conserving water. It allowed the cultivation of a wide range of crops such as vegetables, fruits, and ornamental plants in water-scarce areas.

This technology has not only transformed Israeli agriculture but has also become a global standard, with countries worldwide adopting drip irrigation for its efficiency and sustainability. Israeli companies have become major exporters of drip irrigation systems and expertise, contributing to water-saving practices in regions facing similar environmental constraints. The success of drip irrigation in Israel exemplifies how innovation, combined with a focus on resource conservation, can overcome environmental limitations and make agriculture more sustainable and productive in challenging conditions.

Despite its advantages, its adoption in India has been somewhat limited and half-hearted due to various challenges. One major obstacle is the high initial cost of installation and maintenance, which can be prohibitive for small and marginal farmers. To promote full-scale adoption, government support through subsidies and extension programs is crucial, alongside providing training and easy access to credit facilities. Investing in research to develop cost-effective, durable, and low-maintenance systems suited for local conditions can also help. Demonstration projects showcasing successful implementation can encourage more farmers to switch. When fully adopted, drip irrigation can save up to 60-70% of water, increase crop productivity, enable efficient fertiliser use through fertigation, and reduce weed growth and soil erosion, making it a vital tool for sustainable agriculture in India.

2. Wilmar International in Southeast Asia – Founded in 1991 and headquartered in Singapore, Wilmar International is ranked amongst the largest listed companies by market capitalisation on the Singapore Exchange. Wilmar emphasises sustainability and value addition, exporting high-quality products globally and promoting responsible farming practices. Wilmar has extensive palm oil plantations primarily in Indonesia and Malaysia, which are two of the world's largest producers of palm oil.

At the core of Wilmar's strategy is an integrated agribusiness model that encompasses the entire value chain of the agricultural commodity business, from origination to processing, branding, merchandising and distribution of a wide range of edible food and industrial products. The Group's business activities include oil palm cultivation, oilseed crushing, edible oils refining, flour and rice milling, sugar milling and refining, manufacturing of consumer products, ready-to-eat meals, central kitchen products, speciality fats, oleochemicals, biodiesel and fertilisers as well as food park operations.

Wilmar has over 1,000 manufacturing plants and an extensive distribution network covering China, India, Indonesia and some

50 other countries and regions. Through scale, integration and the logistical advantages of its business model, Wilmar is able to extract margins at every step of the value chain, thereby reaping operational synergies and cost efficiencies. Founded in 1999, selling Fortune brand edible oils, the India venture was initially a 50-50 joint venture between Wilmar International and the Adani Group. From February 2025, it has become a subsidiary of Wilmar International.

Currently, India faces a shortfall of approximately 14-15 million tonnes annually in meeting its domestic edible oil requirements. The country's annual consumption of edible oils exceeds 25-26 million tonnes, while production from local sources such as groundnut, mustard, and soybean accounts for only about 10 to 12 million tonnes. Businesses like Wilmar International can play a crucial role in bridging this gap by leveraging their extensive global experience, supply chain infrastructure, and sourcing capabilities. They can contribute to stabilising prices and reducing shortages by increasing local processing capacity and creating strategic partnerships with Indian farmers and suppliers. Such multinational firms can bring in technology, investment, and operational efficiencies that are essential to meet India's growing demand for edible oils, supporting both food security and economic growth in the sector.

3. Ethiopian Coffee Exporters – Ethiopia is considered the birthplace of coffee, with a rich heritage dating back centuries, and it is famous for its high-quality Arabica beans. Coffee is a cornerstone of Ethiopia's economy and cultural identity, with the coffee sector providing livelihoods for millions of farmers in rural areas.

Ethiopian coffee exporters work closely with smallholder farmers. Many Ethiopian coffee cooperatives now process and package their coffee beans locally, adding value and branding their products as premium, ethically-sourced coffee. This boosts farmers' incomes and enhances Ethiopia's reputation in the global market. Ethiopian coffee exporters play a vital role in one of Africa's most renowned coffee-producing countries. Cultivating coffee on small plots of land using traditional methods, the exporters focus on preserving the unique flavours and qualities of Ethiopian coffee, which are highly valued on the international market. They export a variety of coffee types, each with distinct flavour profiles.

Kaffa, in south-west Ethiopia, is the natural habitat to all other Arabica coffee plants. Not cultivated nor maintained, the coffee grows wild in the middle of deep forests. Sidamo coffee is well-balanced with cupping notes exhibiting berries and citrus with complex acidity. Grown in the Sidamo highlands at elevations from 1,500 up to 2,200 meters, this coffee has more time to absorb nutrients and develop more robust flavours based on the local climate and soil conditions. Sidamo coffee includes Yirgachefe Coffee and Guji Coffee. Guraferda features a small and greyish bean, yet is valued for its deep, spicy and wine or chocolate-like taste and floral aroma. Harar comes from the Eastern highlands of Ethiopia and is known for its distinctive fruity, wine flavour.

India lacks a strong, unique branding on a global scale. One primary factor is the country's vast diversity in culture, language, and regional identities, which makes it challenging to create a single, unified brand image that represents the entire nation. Additionally, India has historically faced challenges related to infrastructure, quality standards, and consistent marketing strategies, which are essential for building a recognisable global brand. The focus has often been on exporting raw materials or products rather than investing in branding and value addition, limiting the development of an iconic Indian identity in international markets.

Located in the Eastern Ghats of Andhra Pradesh, Araku Valley's unique climate, rich volcanic soil, and shade-grown coffee cultivation contribute to the distinct flavour profile of its coffee. The Arabica Typica and Arabica Bourbon varieties are mostly cultivated. The region practices sustainable and organic farming methods, which not only enhance the quality of the coffee but also promote eco-friendly agriculture and support local communities. Araku coffee, promoted through global branding by Naandi Foundation and international collaborations, is often described as aromatic, mild, and flavourful, making it highly sought after in both domestic and international markets. I envision Araku Arabica coffee becoming a globally renowned top premium product both domestically and internationally.

4. Indigo Agriculture in the United States – Indigo Agriculture is an innovative American company focused on sustainable farming and agriculture technology. Founded in 2014, Indigo aims to improve crop production, environmental health, and farm profitability through the integration of biology, data analytics, and digital tools. Utilising microbial technology and digital tools, Indigo works on sustainable farming practices that improve crop health and yield while reducing chemical inputs. Their innovative approach promotes environmental sustainability and crop resilience.

One of Indigo's core initiatives is the development of microbial seed coatings and soil inoculants that enhance plant health and resilience. These natural solutions help crops withstand stress from drought, pests, and diseases, reducing the need for chemical inputs and promoting environmentally sustainable practices. The company's mission is to harness beneficial microbes and biologicals to create more productive and sustainable agricultural systems. In addition to biological innovations, Indigo employs data-driven approaches such as digital platforms that connect farmers, provide real-time insights, and optimise resource use. This technology enables farmers to make more precise decisions on irrigation, fertilisation, and crop management, leading to increased yields and reduced environmental impact.

Indigo also offers a marketplace for farmers to sell their crops directly to buyers, creating more transparency and value for agricultural products. Its focus on sustainability, innovation, and digital integration exemplifies a modern agribusiness model that aims to align economic growth with environmental stewardship. Overall, Indigo Agriculture's use of biological products and digital tools positions it as a leader in the 21st-century sustainable farming movement, contributing to healthier soils, more resilient crops, and a more efficient agricultural ecosystem.

Other notable examples of microbial seed coatings and soil inoculants include mycorrhizal fungi products. These inoculants contain beneficial fungi that form symbiotic relationships with plant roots, helping them absorb nutrients more efficiently, particularly phosphorus. Companies like Mycorrhizal Applications (MA) in Grants Pass, Oregon and Bio Nova International, Waalwijk in the southern Netherlands, develop these products to promote healthier root development, increase crop yields, and improve soil health, especially in degraded or nutrient-poor soils.

Rhizobium inoculants are beneficial soil microbes used primarily for leguminous crops like beans and peas. These bacteria form nodules on plant roots and fix atmospheric nitrogen into a form the plant can absorb, reducing dependence on chemical fertilisers. Examples include products like Legumefix and Rhizogrow, which enhance plant growth, especially in sandy soils. Plant growth-promoting rhizobacteria (PGPR) boost plant health by producing growth hormones, solubilising phosphates, and suppressing soil pathogens, leading to higher yields. Vesicular-arbuscular mycorrhizal fungi (VAMF) form networks in the soil, improving water and nutrient uptake—particularly phosphorus—and protecting plants from soilborne diseases.

Advanced biofertilisers now incorporate genetically optimised microbes and microbial consortia—combinations of beneficial bacteria and fungi—that work synergistically to enhance nutrient efficiency, disease resistance, and stress tolerance. For instance, pairing nitrogen-fixing bacteria with VAMF maximises nutrient absorption and resilience. Innovations also include nanotechnology-based delivery systems that improve microbial stability and efficacy. Organic biostimulants, such as seaweed extracts and humic acids, further support plant growth by enhancing nutrient uptake and stress tolerance naturally. These biostimulants are eco-friendly options enhancing soil health and crop productivity, playing a vital role in sustainable agriculture. Overall, these biological solutions contribute to resilient, productive, and environmentally friendly farming systems, aligning with modern sustainable practices.

A possible biostimulants revolution could significantly transform agriculture by promoting sustainable, resilient, and more productive farming systems. It would enable farmers to improve crop yields and quality naturally, reducing reliance on chemical fertilisers and pesticides, which can degrade soil health and harm the environment. Biostimulants enhance nutrient uptake, improve plant resilience to stresses such as drought, salinity, and pests, and promote healthier soil ecosystems through microbial activity. This shift toward biological solutions would support eco-friendly farming practices, contribute to climate resilience, and help address soil degradation and resource depletion. Overall, a biostimulants revolution would lead to more sustainable, efficient, and environmentally friendly agriculture, ensuring food security for the growing global population while safeguarding natural resources.

India has significant potential to lead the biostimulants revolution in the global south due to its vast agricultural sector, rich biodiversity, and growing emphasis on sustainable farming practices. The country's extensive agricultural knowledge base, combined with innovative research and development in biostimulants—such as organic extracts, microbial inoculants, and advanced formulations—positions India as a key player. Additionally, the increasing demand for eco-friendly and organic products, along with government initiatives promoting sustainable agriculture, further supports this potential. India's large agricultural market, coupled with its capacity for manufacturing and exporting biostimulants, can help transform farming across many developing countries in Asia, Africa, and Latin America, promoting higher yields, better soil health, and resilience against climate stresses.

I envision a "Brown Revolution" bringing in a transformative shift in agriculture towards more sustainable, eco-friendly, and soil-centric farming practices. Unlike the Green Revolution, which focused primarily on increasing crop yields through synthetic fertilisers, pesticides, and high-yield variety seeds, the Brown Revolution emphasises restoring and enhancing soil health, promoting organic growth, and prioritising environmental sustainability. This revolution aims to address the growing concerns over soil degradation, water pollution, and climate change caused by chemical-intensive agriculture by promoting practices that are compatible with natural ecosystems.

Central to the Brown Revolution is the idea of revitalising soil health. Healthy soil is the foundation of productive and resilient agriculture, but over decades of chemical use, soils have become depleted of organic matter, biodiversity, and essential nutrients. The Brown Revolution seeks to reverse this trend by encouraging organic inputs such as compost, organic fertilisers, and bio-inoculants like microbial solutions and biostimulants. These natural amendments improve soil structure, increase microbial diversity, and enhance nutrient cycling, leading to better water retention, reduced erosion, and higher crop resilience against pests and droughts.

Furthermore, the Brown Revolution champions organic growth—cultivating crops in harmony with nature, reducing or eliminating synthetic chemicals, and adopting eco-friendly farming techniques such as crop rotation, cover cropping, and integrated pest management. Such practices not only improve the sustainability of farming systems but also safeguard biodiversity, pollinator health, and overall ecosystem balance. This approach also results in healthier foods, free from harmful chemical residues, thereby benefiting consumer health.

A pivotal goal of the Brown Revolution is to shift from dependence on chemical inputs towards biologically driven solutions. This includes the widespread use of biofertilisers, biostimulants, and beneficial microbes like Rhizobium and mycorrhizal fungi, which naturally enhance plant growth, improve nutrient uptake, and bolster resistance to biotic and abiotic stresses. By fostering a more resilient and sustainable agricultural system, the Brown Revolution aims to reduce environmental pollution, conserve water resources, and mitigate climate impacts.

In essence, the Brown Revolution envisions an agriculture that restores natural soil fertility, protects ecosystems, and promotes resilience against climate variability. It advocates for a balanced approach—integrating traditional ecological knowledge with modern scientific innovations—focused on nurturing the land and supporting farmers' livelihoods. This revolution can help create a sustainable agricultural future where productivity is decoupled from environmental harm, ensuring food security while preserving the planet for generations to come.

•••

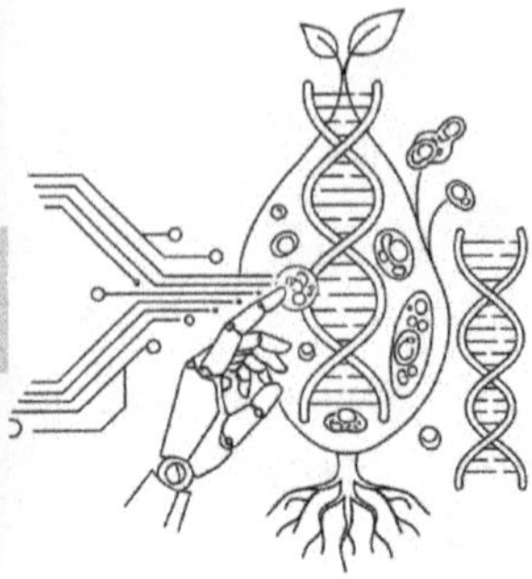

9

BIOPLASTICS

The age of fossil-fuel plastics is drawing to a close; in its place rises a new generation of degradable, biodegradable materials — where technology turns crops, microbes, and waste into polymers that serve without scarring the earth.

Chile is often regarded as a progressive economy and a peaceful, modern country due to its stable political environment, strong economic growth, and commitment to social development. Over the past decades, Chile has transformed into one of Latin America's most economically developed nations by implementing sound economic policies focused on free markets, attracting foreign investment, and promoting innovation. This approach has resulted in steady GDP growth, low inflation, and a relatively high standard of living for many Chileans. But in the context of this book, Chile is actively involved in the development and promotion of bioplastics as part of its broader strategy to enhance sustainability and reduce plastic pollution. The country's rich natural resources, including agricultural products like potatoes, grapes, and various fruits, provide raw materials for bioplastic production.

Several Chilean companies and research institutions are working on creating bioplastics from local biomass, aiming to develop eco-friendly alternatives to conventional plastics. These efforts include producing biodegradable packaging, agricultural films, and disposable items that degrade faster and have less environmental impact. Moreover, Chile's commitment to sustainability is reflected in its policies encouraging the circular economy, waste reduction, and renewable materials. The country's diverse ecosystems and agricultural sectors offer opportunities to scale bioplastic production sustainably, aligning with global trends toward reducing dependence on fossil fuels and minimising plastic waste.

The term "circular economy" began to gain prominence in the early 2010s, although the underlying concepts date back much earlier. It is often associated with efforts to design out waste and keep products and materials in use for as long as possible, promoting sustainability. The Ellen MacArthur Foundation, established in 2010, has been pivotal in popularising the term and concept globally. Ellen MacArthur is a renowned British sailor who became famous for her achievements in solo ocean racing. After retiring from sailing, she dedicated herself to promoting sustainability and environmental awareness. Chile has shown interest in promoting the circular economy and is the first to integrate circular economy principles into its policies.

Interestingly, while Chile is home to the Atacama Desert, the driest desert in the world, it also has the stunning Patagonian region, glaciers, mountains, and fjords — deep, glacially carved inlets of the sea that are surrounded by steep cliffs or mountains. Chile has experienced both the benefits and challenges associated with mining. The country is one of the world's largest producers of copper, silver, and other minerals, which are a significant part of its economy. However, mining caused deforestation, water pollution, and soil degradation, affecting local communities and leading to conflicts. The Mapuche people, native to south-central Chile and southwestern Argentina, resisted mining. So, Chile was already prepared to lap up the idea of a circular economy, which emphasises the importance of giving back when taking from the Earth.

Bioplastics are one of the promising ideas within the circular economy framework. They offer an environmentally friendly alternative to traditional plastics derived from fossil fuels, as bioplastics are made from renewable biological sources such as corn (*Zea mays* L.), sugarcane (*Saccharum officinarum* L.), or algae. Bioplastics can be part of a sustainable life cycle when properly managed through recycling, composting, or biodegradation systems. As part of a broader shift toward sustainable materials and waste reduction, bioplastics hold significant potential for advancing circular economy practices.

The development of plastics in the early 20th century revolutionised manufacturing because of their versatility, durability, and lightweight nature. Plastics became mainstream due to a combination of technological, economic, and societal factors that made them highly attractive across various industries. One key factor was the invention and commercialisation of synthetic plastics like Bakelite, polyvinyl chloride (PVC), and polyethene, which provided inexpensive alternatives to traditional materials such as glass, wood, and metal. These plastics could be easily moulded into a wide array of shapes and sizes, enabling innovations in packaging, construction, automotive, electronics, and consumer goods.

Economically, the mass production capacity and chemical advances allowed plastics to be produced at scale and at low costs, making them widely accessible to both manufacturers and consumers. Their durability and chemical resistance extended product lifespans, further boosting their use. Society's adoption of plastics was accelerated by their convenience, such as in packaging (e.g., plastic wraps, bottles), and their role in everyday products. After World War II, plastics-powered consumer culture expanded rapidly, with inexpensive, disposable items becoming commonplace. Over time, plastics' utility, affordability, and adaptability led them to become integrated into daily life worldwide, turning them into a dominant material of the modern age.

Plastic, unlike organic waste, is not degradable and poses significant hazards to the environment, human health, and ecosystems. Because these plastics do not break down easily in nature, they accumulate over time, causing long-term pollution issues. One of the primary environmental hazards is plastic pollution in oceans, rivers, and landfills. Non-degradable plastics can persist for hundreds of years, harming marine and terrestrial wildlife through ingestion or entanglement. Marine animals often mistake plastics for food, leading to injury or death, and toxins from plastics can enter the food chain, affecting larger species and humans. Additionally, these plastics release harmful chemicals during degradation or through leaching, such as microplastics and toxic additives like phthalates and heavy metals. Even more dangerous, microplastics have been found in drinking water, seafood, and even air, raising concerns about potential health impacts on humans. Their long lifespan means that plastic waste remains in the environment for centuries, causing persistent pollution problems.

India is a powerhouse in the plastics industry, with a significant production capacity. Initially considered a breakthrough, plastic now poses environmental challenges, impacting marine life and land resources. It has become one of the most produced materials and has been an environmental concern for many years. Despite technological advancements, managing plastic waste effectively remains difficult. Bio-based plastics are polymers derived from renewable biomass sources such as sugarcane, starch, plant oils and cellulose. Bioplastics act as an alternative for almost every conventional plastic material and corresponding application.

Bioplastics are a promising sustainable alternative to customary plastics derived from fossil fuels. These materials are produced from renewable biological sources such as corn, sugarcane, and cellulose, which significantly reduces dependence on finite resources. They can be classified into bio-based plastics, which are made from renewable sources, and biodegradable plastics, which can be broken down by microorganisms into natural substances like water, carbon dioxide, and biomass. Some bioplastics are both bio-based and biodegradable, offering an ideal combination for environmentally friendly applications.

Bioplastics can be broadly classified into three major categories. First are bio-based, non-biodegradable plastics. Let us call this BP1. These bioplastics are biologically derived but are not biodegradable. Bio-based bioplastics are 100% mechanically recyclable through current recycling methods. In other words, they can be mechanically recycled in accordance with existing recycling streams. These materials possess properties that are identical to their conventional versions (PE, PP, PET, PA), and they can help to reduce a product's carbon footprint through recycling.

Bio-based polyethene terephthalate (Bio-PET) is similar in properties to conventional PET used in bottles and textile fibres, but is partially made from renewable raw materials like sugarcane. These are used in foils/bags, bottles, clothing, textiles and other packaging. Since PET is the largest polyester used for fibres (65%), bio-PET offers a bio-based alternative in this sector. Bio-based polyethene (Bio-PE) is similar to its fossil-based counterpart but produced from ethanol, derived from crops like sugarcane or corn. These are used in many applications such as thin films, packaging, plastic containers, bottles, bags, plastic toys, wire and cable insulations, medical tubing, and injection moulding, among others. Bio-based polyamides/nylons (Bio-PA) produced from vegetable oil retains the desirable properties, such as flexible to rigid and food safe. These are used in a wide range of industrial applications in sectors including automotive, electrical, textile and medical industries.

Second are bio-based, biodegradable plastics, which we will call BP2. These bioplastics are made from renewable biological material sources, such as starch, cellulose and proteins. Some of these materials biodegrade predominantly under composting conditions, while others are in less controlled environments. PLA is a biodegradable and recyclable polyester made from renewable feedstock. Lactic acid, as the raw material, is produced by fermentation of glucose or sucrose and is refined to a high purity. PLA is better than other plastics in thermostability, and it is the best in terms of 'hardness', 'mechanical strength' and 'transmittance'.

As per production cost, PLA has the lowest production cost of rigid materials. Polyhydroxyalkanoates (PHA) are biodegradable and are extracted from bacterial fermentation processes. PHA bioplastic is good for film forming and hydrolysis resistance, and is therefore best suited for packaging water, but is also used for general packaging purposes. Polybutylene succinate (PBS) is one of the most important biodegradable polyesters synthesised by polycondensation between succinic acid and butanediol. Its high flexibility guarantees its use in many applications involving film production. Biodegradable/compostable starch blends are fully biodegradable plastics made from starch-rich plants like potatoes, wheat or corn. Starch-based bioplastics can be used to create agricultural mulch films, which help suppress weeds and retain moisture in the soil. With the ability to blend with other polymers, starch-based bioplastics are commonly used in packaging materials such as trays, loose-fill foams, cups, plates, and cutlery. They can also be used to create biodegradable films for food packaging.

The third type, fossil-based biodegradable plastics (BP3), are biodegradable plastics produced from fossil-based resources, such as PBAT and PCL. Polybutylene Adipate Terephthalate (PBAT) is a type of thermoplastic polymer known for its biodegradability. It is derived from common petrochemicals—purified terephthalic acid (PTA), butanediol and adipic acid. PBAT's hydrolysis resistance makes it the best material to package food and other snacks/cereals. It is also used for the production of shopping and express bags, packaging films and 3D printing material. Polycaprolactone (PCL) is a synthetic, semi-crystalline, biodegradable polyester with a melting point of only 60 °C, which means it can be shaped by hand after a few minutes in hot water. PCL has numerous applications in packaging, scaffolds, prosthetics, sutures, drug delivery, films, carry bags, pouches, trays, reusable dishes, membranes and other fields.

The production of bioplastics involves various biotechnological and chemical processes. For example, PLA is created through the fermentation of carbohydrate-rich feedstocks to produce lactic acid, which is then polymerised into plastic. Similarly, PHAs are synthesised by bacteria during microbial fermentation of organic materials. Advances in genetic engineering and fermentation technology continue to improve the efficiency and sustainability of bioplastic manufacturing, making these materials more accessible and cost-effective. Let's dive deeper.

PLA is synthesised from lactic acid, which is produced via fermentation of carbohydrates from crops like corn, sugarcane, or cassava. The lactic acid molecules are then polymerised to form PLA, a clear, biodegradable thermoplastic used in packaging, disposable tableware, and medical implants. As a leading bioplastic, PLA plays a vital role in advancing sustainable materials. With major manufacturers like American company NatureWorks and Dutch Corbion innovating in production and expanding their product lines, PLA continues to grow as a versatile and eco-friendly alternative across multiple industries.

Another superstar bioplastic is Bio-PE, also known as bio-based polyethene. It is made from renewable resources like sugarcane or corn, instead of traditional fossil fuels. It's considered a more sustainable alternative to conventional polyethene (PE) because it reduces reliance on fossil fuels and can lower the carbon footprint of plastic production. Bio-PE shares the same chemical, physical, and mechanical properties as conventional PE, making it a "drop-in" replacement. This means it can be used in the same applications and recycled using existing PE recycling infrastructure.

In another big shift, industry labs and now academic institutions are leading bioplastic research. Leading manufacturers like Braskem in Brazil, Dow and ExxonMobil in the United States, and Badische Anilin und Sodafabrik (BASF) in Germany are advancing the commercialisation of Bio-PE, making it a viable, sustainable alternative for various plastic applications. Its compatibility with existing production infrastructure and identical properties to conventional PE facilitate its adoption across industries such as packaging, foodservice, and agriculture, contributing to a more sustainable future. Aloke Lohia, born in 1958 in Kolkata, founded Indorama Ventures in Thailand in 1994.

It is now among the world's Top 30 chemical companies and a leading global manufacturer of polyethene, including bio-based variants, operating facilities in India.

PHAs are another class of bio-based plastics generated by bacteria during microbial fermentation of organic substrates. PHAs are biodegradable and have properties similar to polypropylene or polyethene, but they decompose naturally in soil and marine environments. They are used in applications such as biodegradable packaging, agricultural films, and medical devices, and their properties can be tailored by varying the monomer composition, enabling a range of applications from flexible films to rigid plastics. Danimer Scientific, United States; Medipha, France; BIO-On, Italy; Kaneka Corporation and Mitsui Chemicals, Japan are advancing the PHA technology to play a crucial role in creating a circular, environmentally responsible plastics industry.

Cellulose-based plastics are derived from natural cellulose, the most abundant organic polymer found in plant cell walls. These plastics are produced by chemically modifying cellulose, resulting in materials such as cellulose acetate, cellulose nitrate, and regenerated cellulose. Known for their biodegradability, transparency, and renewable origin, cellulose plastics are widely used in textiles, packaging, and photographic films.

Prominent manufacturers are Eastman Chemical Company and Celanese Corporation, United States; Sateri, China; and Lenzing AG, Austria. In India, JK Paper produces regenerated cellulose films and speciality cellulose products used in packaging and industrial applications. As demand for biodegradable and eco-friendly materials grows, cellulose-based plastics are positioned to play a vital role in the transition toward sustainable materials management.

Hemp (*Cannabis sativa* L.) is connected to bioplastics as a source for producing biodegradable, renewable materials derived from its high cellulose content. Unlike its cousin marijuana, industrial hemp contains very low levels of tetrahydrocannabinol (THC), the psychoactive compound, making it suitable for legal cultivation and a variety of commercial uses. Hemp bioplastics offer environmental advantages, including reduced reliance on fossil fuels, lower carbon footprints, and natural biodegradability, positioning hemp as a sustainable alternative to petroleum-based plastics for products ranging from packaging to automotive parts.

Hemp fibres are strong, durable, and biodegradable, making them ideal for clothing, shoes, and accessories. Hemp textiles are often marketed as sustainable alternatives to cotton and synthetic fibres. Hemp hurds and fibres are used to produce biodegradable plastics, insulation, hempcrete (a lightweight, insulating concrete alternative), and flooring materials. These eco-friendly building products are gaining popularity in green construction projects.

The legalisation of hemp cultivation in many countries, especially following the 2018 Farm Bill in the United States, has spurred rapid industry expansion. Canada, China, and various European nations are also major players, producing hemp for both domestic use and export. The hemp industry is poised for continued growth, driven by sustainability trends, innovation in products, and expanding legal acceptance. Advances in breeding, processing, and product development will likely unlock more applications. The industry holds promise not only in improving environmental sustainability but also in fostering economic growth through rural development and job creation.

As a renewable, biodegradable, and multipurpose crop, hemp stands at the forefront of sustainable agriculture and industry. Its diverse applications—from textiles to health products—highlight its potential to contribute significantly to greener, healthier economies worldwide. While initially constrained by stringent drug laws, India is gradually embracing hemp cultivation through a combination of state initiatives and amended regulations. With careful regulation, the country can tap into hemp's economic, environmental, and medicinal benefits, fostering a sustainable and innovative industry. In 2019, the Central Government of India reclassified hemp in the Horticulture Policy and other related regulations to promote its cultivation. Several Indian states, including Uttarakhand, Himachal Pradesh, and Uttar Pradesh, issued licenses and established policies to promote hemp cultivation for industrial use. India's hemp industry shows promising growth, bolstered by progressive state policies, the global trend towards sustainable materials, and increasing awareness of hemp's benefits.

Where do we go from here? The latest BioE3 (Biotechnology for Economy, Environment and Employment) policy aims to foster high-performance bio-manufacturing in India. The said policy has been launched with the aim of facilitating a shift from chemical-based industries to sustainable bio-based models, promoting a circular bioeconomy and working towards achieving net-zero carbon emissions. Further, the policy recognises biopolymers as one of the strategic sectors for sustainable growth, which aligns with India's vision of a cleaner, greener, and more prosperous future. To address environmental concerns associated with conventional plastics and the escalating issue of plastic pollution, the transition to bio-based plastic alternatives that support a circular economy presents a promising and sustainable approach.

India has been active in combating plastic pollution with a series of legislative and innovative measures for several decades, since the introduction of plastics. India's Climate Action Plan 2008 envisages reducing carbon emissions by 1 billion tonnes by 2030; reducing carbon intensity in the economy by 45% by 2030 over 2005 levels; achieving the target of net-zero emissions by 2070. Swachh Bharat Mission – Urban 2.0 focuses on source segregation of solid waste, utilising the principles of reduce, reuse, recycle, and scientific processing of all types of municipal solid waste. Circular Economy Roadmap 2021 aims to foster research and industry partnerships between India and Australia and co-develop a roadmap for India's transition to a circular economy in the plastics sector.

Superseding the earlier Plastic Waste (Management and Handling) Rules, 2011, the Ministry of Environment, Forest and Climate Change (MoEFCC) notified the Plastic Waste Management Rules, 2016. Producers, importers and generators of plastic waste were mandated to implement a collect-back system through the concept of extended producer responsibility (EPR). In no uncertain terms, it is declared that biodegradable plastics are those which undergo degradation by biological processes in a specific environment, such as soil, landfill, sewage sludge, fresh water, marine environments, without leaving any microplastics or visible, distinguishable or toxic residue. Compostable plastics mean plastic that undergoes degradation by biological processes during composting to yield CO_2, water, inorganic compounds, and biomass at a rate consistent with other known compostable materials, excluding conventional Petro-based plastics, and does not leave visible, distinguishable, or toxic residue. Compostable plastics can also be anaerobically digested to produce biogas and nutrient-rich manure. On July 1, 2022, India banned the manufacture, import, and use of certain identified single-use plastic items, targeting the adverse environmental, economic, and social ramifications of such materials.

The Department of Chemicals and Petrochemicals has launched a Scheme for setting up 'Centres of Excellence' in biopolymers and eco-friendly polymers. refining recycling technologies, and advancing innovative methods for the collection, sorting, purification and production of recycled items, among other objectives. Key centres include the Centre of Excellence in Sustainable Polymers (CoE-SusPol) at IIT Guwahati, the Parimal and Pramod Chaudhari Centre of Excellence and Innovation (CoEI) for Biopolymers by Praj Industries, Pune, and Institute of Chemical Technology (ICT) in Mumbai, the CSIR-National Chemical Laboratory (CSIR-NCL) in Pune and the Central Institute of Plastics Engineering and Technology (CIPET) in Chennai and Bhubaneswar. The Tripura State Pollution Control Board has released a report titled "Bioplastics: An Alternative to Conventional Plastics," which underscores the need for regulatory support to boost bioplastics production. However, there is no specific scheme yet for incentivising bioplastic production and disposal in the country.

Thailand offers a great example that we must emulate. In 2023, Thailand became the second-largest bioplastics producer with a capacity of 95,000 tonnes, following the capacity of 1,50,000 tonnes of the USA. If we compare the sizes of the two economies, it is a great achievement that highlights the country's commitment to sustainability and environmental stewardship. It happened because Braskem, the world's largest biopolymer producer, along with SCG Chemicals Plc (SCGC), has built a 2,00,000 tonnes per year bio-based polyethene plant in Thailand. Merely law-making will not help; we have to build capacity.

Supporting the circular economy in bioplastics relies heavily on developing clusters, a long-established concept. Throughout history, markets like Sarafa Bazaar for bullion, Bajaja Bazaar for textiles, Dal Mandi for pulses, and Anaaj Mandi for grains have exemplified how sectors thrive within interconnected ecosystems. In Kota, Rajasthan, coaching centres have formed a cluster, illustrating how different sectors prosper collectively. Customers benefit from the availability of multiple sellers in a single location, while support services such as labour and transportation are pooled, creating a vibrant ecosystem.

The Government of India has implemented various cluster schemes to foster similar ecosystems that promote innovation, skill development, investment, and employment. For instance, the PM MITRA Scheme has bolstered the Indian textile industry by enabling large-scale operations and reducing logistics costs through centralised facilities that house the entire value chain. The Pradhan Mantri Matsya Sampada Yojana (PMMSY) has transformed aquaculture by developing hatcheries, grow-out ponds, fish markets, feed mills, storage units, and quality testing laboratories. Additionally, the Mega Food Parks Scheme (MFPS) offers capital grants of up to 75% to establish processing infrastructure near farms, enhancing transportation, logistics, and centralised processing.

The Micro and Small Enterprises – Cluster Development Program (MSE-CDP) aims to enhance productivity and competitiveness by providing government financial support for establishing Common Facility Centres (CFCs) in existing clusters and for upgrading or creating Industrial Areas, Estates, and Flatted Factory Complexes. For constructing tangible assets such as processing centres, testing facilities, design centres, and plug-and-play infrastructure, the government covers up to 80% of the project costs. Additionally, development activities like drainage, power distribution, roads, and other infrastructure in existing or new industrial estates or complexes are financed up to 70% of the project costs. While there are upper limits, such comprehensive support has rarely been available before.

In the plastics sector, a dedicated scheme exists to develop infrastructure and common facilities to help the industry move up the value chain and contribute more effectively to the economy. The Central Government provides grants covering up to 50% of project costs, with the remaining funding contributed by the State Government, State Industrial Development Corporations, or similar agencies.

As the world increasingly seeks alternatives to conventional plastics and the global bioplastic industry expands—with many countries incentivising companies to establish bioplastic manufacturing—India has a significant opportunity to strengthen its presence in the global value chain. This can be achieved by creating demand for bioplastics and promoting domestic production through strategic incentives. With appropriate support, the bioplastics sector has the potential for rapid and exponential growth.

Despite ongoing government efforts to combat plastic pollution, a dedicated focus on bioplastics policies is essential. Learning from global best practices, particularly from countries like Chile and Thailand, India can nurture a thriving bioplastics industry. Developing this sector offers clear benefits, including reducing India's dependence on importing bioplastics, creating new employment opportunities through the establishment of manufacturing plants, and increasing income for farmers by utilising agricultural produce such as sugarcane for bioplastic production.

In a nutshell, a comprehensive 'National Bioplastics Policy' will play a vital role in tackling the rising challenge of conventional plastic waste, supporting the government's commitment to sustainability goals, including achieving 'Net Zero' by 2070. Implementing well-designed incentive structures and subsidies can help offset higher production costs and promote widespread adoption of bioplastics. This strategic approach will not only improve India's Environmental Performance Index (EPI) but also strengthen the country's long-term economic and environmental resilience.

The creation of 'bioplastics clusters', focused on advancing the product recovery phase of the end-of-life value chain, will help embed industry practices within a circular economy framework. To develop these clusters effectively, the government must provide holistic support across several key areas. First, enhancing knowledge and technology through funding research and development, fostering partnerships between academia and industry, and incentivising the adoption of innovative bioplastic technologies are essential. Second, investing in critical infrastructure for the production, processing, and distribution of bioplastics is necessary to ensure facilities are environmentally sustainable and economically feasible. Lastly, promoting job creation through policies that encourage industry growth—including workforce training programs and tax benefits for companies generating new employment—will be crucial for creating a vibrant, sustainable bioplastics industry.

The development of bioplastics clusters promises significant advantages, benefiting not only leading industry players involved in bioplastics manufacturing but also micro, small, and medium enterprises (MSMEs) such as converters, compounders, and others integral to the entire bioplastics value chain. By focusing on these areas, the central government can accelerate industry growth, foster innovation, promote sustainability, and strengthen the overall economy.

However, fostering know-how and innovation through collaboration among all stakeholders in the value chain requires a comprehensive, interdisciplinary approach. This involves establishing research laboratories, centres of excellence, testing and design facilities, and facilitating strong links with universities and government agencies. Engaging all partners across the entire value chain is essential.

India has the potential to integrate bioplastics into broader environmental and economic strategies by leveraging its biomass resources and leadership in sugar production. Creating supportive policy frameworks, providing incentives for bioplastics manufacturing, and adopting a cluster-based development approach are crucial steps toward establishing India as a resilient circular economy hub.

•••

10

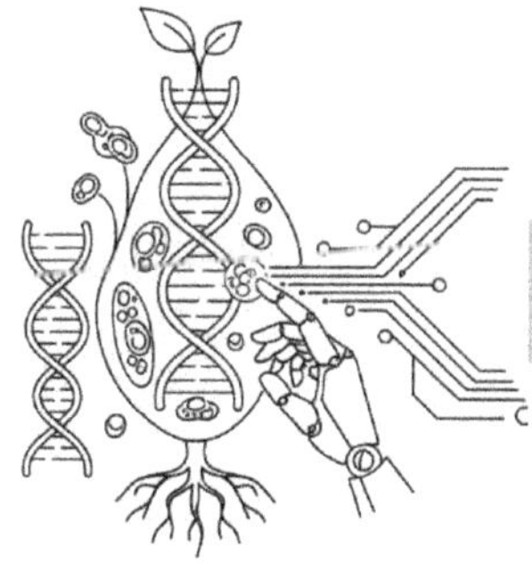

BIOENERGY

Energy is life distilled — and when drawn from living cycles, it lights not only homes but the horizon of generations to come.

Bioenergy in plants is primarily stored in the form of chemical compounds, including starch, sugars, and lipids. Starch, a polysaccharide composed of glucose units, is primarily stored in roots, seeds, and tubers, serving as a short-term energy reserve that can be broken down into glucose for energy when needed. Sugars, such as sucrose, are transported through the plant's phloem to various parts for immediate use or storage. Lipids, including fats and oils, are stored primarily in seeds and other tissues as long-term energy reserves. These compounds store energy in their chemical bonds, which can be released through metabolic processes, such as respiration, to power the plant's growth and development.

When biomass is burned or undergoes biological conversion processes, such as fermentation or anaerobic digestion, these chemical bonds are broken down, releasing the stored energy in the form of heat, light, or biofuels. For example, combustion of biomass converts the chemical energy into thermal energy, which can be used for heating or electricity generation. Biological processes, such as fermentation, produce bioethanol and biogas, which can be utilised as renewable fuels. Overall, biomass energy is accessible by converting the stored chemical energy within organic materials into usable forms through combustion or biochemical methods.

The idea of making ethanol from biomass has a long history, with its roots dating back to the 19th century. However, it gained significant attention and technological development, particularly during the 20th century, especially in response to the oil crises of the 1970s. The first large-scale ethanol production from biomass, mainly from sugarcane and starch-rich crops, began in Brazil in the 1970s, making it a pioneer in bioethanol technology. This period marked the maturation of biomass-based ethanol as a viable renewable fuel source. Since then, advancements have continued, expanding to include cellulosic ethanol from non-food plant materials, with ongoing research to improve efficiency and sustainability.

Brazil's transformation into a bioenergy leader is a remarkable story of strategic innovation, sustainable development, and economic growth. Over the past few decades, Brazil has harnessed its abundant natural resources, particularly its vast land, favourable climate, and renewable biomass, to establish a robust bioenergy sector. This transformation has positioned Brazil as a global pioneer in the production of biofuels, especially ethanol, and in the development of integrated renewable energy systems. The roots of Brazil's bioenergy leadership date back to the 1970s oil crisis, which prompted the country to seek alternatives to reliance on imported fossil fuels. The National Alcohol Program (Proálcool) in 1975 was a crucial turning point.

The initiation of Proálcool was motivated by multiple factors. The 1973 oil embargo by OPEC countries starkly highlighted the vulnerability of nations heavily reliant on oil imports. Brazil's government recognised ethanol from sugarcane as a viable, renewable, and locally available resource that could serve as a clean-burning alternative fuel. President Ernesto Geisel, notable for steering Brazil towards a process called "decompression" or "gradual political liberalisation," which aimed to ease some of the repressive measures of previous military rulers while maintaining control, led it from the front.

Proálcool encompassed several key components: massive investments in sugarcane cultivation, the development of ethanol production facilities, and the promotion of flexible-fuel vehicles that could operate on ethanol, gasoline, or mixtures of both. These vehicles, known as "flex-fuel," became a hallmark of Brazil's fuel strategy. The government also implemented policies such as tax incentives, marketing campaigns, and the creation of a regulated fuel distribution system to encourage widespread adoption. Throughout the 1980s and 1990s, Proálcool faced challenges, including fluctuating oil prices, economic instability, and the global reduction in oil prices, which temporarily diminished the appeal of ethanol. Despite these hurdles, the program persisted and evolved, with technological improvements, increased efficiency, and expanded infrastructure contributing to the sustainable growth of ethanol use.

A significant milestone was Brazil's development of flex-fuel vehicles in the early 2000s, making ethanol a practical and accessible option for consumers. This innovation dramatically increased ethanol consumption, positioning Brazil as the world's leading producer and consumer of bioethanol. The program also contributed to rural development by creating jobs and promoting sustainable agriculture practices. Thereafter, Proálcool remains a vital component of Brazil's energy policy.

Brazil's extensive sugarcane cultivation has been the cornerstone of its bioenergy success. The efficiency of sugarcane as a biomass crop, combined with advances in fermentation technology, enabled Brazil to produce ethanol at competitive prices. Today, Brazilian ethanol is recognised worldwide for its high quality and sustainability standards. The country also pioneered flexible-fuel vehicles that could run on gasoline, ethanol, or any blend of both, fostering a flexible and resilient energy infrastructure. Beyond ethanol, Brazil has diversified its bioenergy portfolio to include biodiesel production from soybean and other feedstocks, as well as bioelectricity generated from biomass waste in sugar mills and forest residues. The integration of these renewable sources has contributed to Brazil's energy security and helped reduce greenhouse gas emissions significantly.

Brazil's commitment to sustainability has been pivotal in its transformation. The country has adopted strict environmental standards for biomass cultivation and biofuel production, ensuring that growth does not come at the expense of deforestation or biodiversity loss. Furthermore, Brazil's emphasis on rural development and job creation through bioenergy initiatives has fostered social inclusivity and economic stability in many regions. Internationally, Brazil's bioenergy expertise has positioned it as a global leader. The country actively shares its technology, experience, and policies through partnerships and knowledge exchange, influencing bioenergy development worldwide. Its success demonstrates how renewable energy can be a driver of economic development, environmental conservation, and energy independence. What can we learn from this?

India's dependency on imported petroleum oil can indeed be likened to the Sisyphus curse, symbolising an ongoing, arduous struggle. Just as Sisyphus in the Greek myth was condemned to endlessly push a boulder uphill only for it to roll back down, India faces the persistent challenge of reducing its heavy reliance on imported oil amidst complex economic, geopolitical, and infrastructural obstacles.

India is one of the world's largest importers of crude oil, with imported oil accounting for a significant portion of its energy consumption. This dependency exposes the country to volatile global oil prices, affecting its economy and trade balance. Efforts to achieve energy security through domestic production, renewable energy investments, and alternative fuels are ongoing but often hindered by technological, financial, and infrastructural hurdles.

Much like Sisyphus's relentless punishment, India's pursuit of energy independence requires continuous, sustained efforts that sometimes seem to make little long-term progress, especially when geopolitical tensions or market fluctuations cause setbacks. Nonetheless, this struggle also fuels innovation in biofuels, offering hope for a future where dependence on imported petroleum diminishes. The beginning of the quest, however, was ominous. The failed experiment with jatropha as a biofuel crop serves as a valuable lesson in the complexities and challenges of renewable energy initiatives.

Hailed as a promising alternative to fossil fuels, jatropha was promoted for its ability to grow on marginal lands, requiring minimal water and maintenance while producing a high yield of oil suitable for biodiesel production. This optimistic outlook led many countries, particularly India, to invest heavily in large-scale cultivation efforts, expecting it to be a game-changer for sustainable energy and rural development. Dr A.P.J. Abdul Kalam also joined the enthusiasm, and wherever he visited as President of India, he promoted jatropha cultivation.

Many jatropha plantations underperformed, failing to produce the expected yields. Its growth was highly dependent on specific climatic and soil conditions, which were often absent on the so-called marginal lands where it was planted. Instead of thriving, jatropha often struggled, leading to economic losses. The crop also required significant water and care, contradicting initial claims that it could grow on neglected lands with minimal inputs, which in turn increased the cultivation costs substantially. As a result, the biofuel produced was often not cost-competitive with fossil fuels or other renewable energy options.

The overenthusiasm and aggressive promotion of jatropha without sufficient scientific validation contributed to the disillusionment. Policymakers and developers had overestimated the potential of jatropha, which led to significant resource misallocation and financial losses. The experience highlights that developing sustainable bioenergy solutions must be grounded in thorough research, realistic expectations, and comprehensive assessments of ecological and economic factors. Although the jatropha experiment has faded, the lessons learned continue to inform current renewable energy strategies, emphasising the importance of evidence-based approaches and sustainability-focused development.

The change came from an unexpected corner. Historically, India has been one of the world's largest producers of sugar, and periods of bumper harvests have often resulted in excess sugar that exceeds domestic demand. This surplus created a favourable opportunity to diversify the utility of sugarcane by converting it into bioethanol, thus providing a new revenue stream for farmers and sugar mills. This shift helped stabilise the sugar industry's prices and provided economic relief to farmers facing low sugar prices. The government actively promoted this approach through policies that incentivised sugar mills to produce ethanol, including offering obtainment commitments from the oil marketing companies and providing subsidies and favourable prices for ethanol. The availability of surplus sugarcane and molasses, combined with supportive government policies, led to a significant increase in ethanol blending programs across several Indian states. Dr R.A. Mashelkar had been supporting research and innovations in biofuels and became its face.

The formulation of the National Biofuel Policy in 2009 was a watershed moment. It set the stage for large-scale ethanol blending and promoted the use of surplus sugarcane and molasses for ethanol production. Rather than a vision, it was a *fait accompli.* The increasing oil import bill put unbearable pressure on India's economy. The policy aimed to encourage crop diversification, incentivise biofuel production, and set clear targets for blending ethanol with petrol, initially aiming for a 10% ethanol blending (E10), which would be gradually increased to E20 by 2025-26. The policy also emphasised the importance of sustainable practices, ensuring that biofuel cultivation did not compromise food security or lead to environmental degradation. It identified suitable feedstocks such as sugarcane, non-edible oils, maize, jatropha, and other agricultural waste, promoting a balanced approach for different regions.

The formal rollout of E10 in India began around 2003-2005, when the government first started actively promoting ethanol-blended petrol (EBP) programs. Oil marketing companies, including Indian Oil Corporation, Bharat Petroleum, and Hindustan Petroleum, launched pilot projects in select cities to demonstrate the feasibility and benefits of ethanol blending. With a clear cost benefit and no technical issues, E10 became widespread, especially in the strong sugar industries of Maharashtra and Tamil Nadu. Uttar Pradesh, though it joined late, contributed to the eventual success of the ethanol blending. But it was not smooth sailing.

The conflict between distilleries and ethanol producers in India primarily arose due to competing interests over feedstock allocation, pricing policies, and regulatory frameworks. Historically, sugar mills and distilleries relied heavily on sugarcane and molasses for their operations, producing alcohol primarily for the production of spirits and other industrial purposes. As the government shifted focus towards promoting ethanol for blending with petrol, tensions emerged over the availability and pricing of raw materials, especially molasses and surplus sugarcane.

To address these conflicts, the Indian government and relevant authorities implemented several measures to create a more balanced and transparent framework. One of the key steps was the formulation of policies that mandated a fixed price for molasses and sugarcane used specifically for ethanol production, ensuring that farmers and sugar mills received fair compensation. The government also introduced long-term procurement contracts with sugar mills to guarantee a steady supply of ethanol, thereby providing stability and encouraging investments in ethanol manufacturing.

Additionally, regulations were put in place to prioritise the allocation of feedstocks for ethanol production based on national energy goals, while protecting the interests of the sugar industry. The government encouraged the establishment of standalone ethanol distilleries separate from traditional sugar mills, reducing competition for raw materials and allowing for more specialisation. The resolution also involved fostering collaboration among stakeholders—farmers, sugar mills, ethanol producers, and government agencies—through dialogue and incentive programs. These measures aimed to mitigate conflicts, promote sustainable feedstock cultivation, and ensure market stability. Over time, with clear policies, pricing mechanisms, and targeted support, the tensions between distilleries and ethanol producers eased, allowing both sectors to grow in a more coordinated manner.

Ethanol blend has a higher octane rating of 108.5, which is more than petrol (84.4), with smooth acceleration. Ethanol absorbs more heat through evaporation, which cools the air temperature in a vehicle's engine. It increases air-fuel mixture density and improves engine performance. However, ethanol has a lower energy density (26 MJ/kg), while pure petrol has a higher energy density (42 MJ/kg). This means a little more ethanol blend would be consumed. Also, vehicles with older versions cannot handle the E20 petrol blend. However, the 2023 Model Year (MY) model aligns with the performance of E20, and companies are now manufacturing compatible vehicles. It also questions the engine's durability. E20-compatible engines with high compression ratios have high power potential.

The journey toward E20 has been marked by systematic scaling. The ethanol blending rate increased from 8.17% in the Ethanol Supply Year (ESY) 2020-21 to 12.06% in 2022-23, reaching 14.6% in 2023-24, and finally achieving the 20% target by 2025-26. The estimated ₹1,25,000 crore from ESY 2014-15 up to July 2025, accruing to sugarcane farmers reflects the substantial income generation and rural prosperity that can be achieved through the promotion of bioethanol and other biofuels. This income supplement not only improves farmers' livelihoods but also incentivises sustainable agricultural practices and crop diversification, reducing reliance solely on traditional sugar production. Simultaneously, savings of more than ₹1,44,000 crore foreign exchange, net CO_2 reduction of about 700 lakh metric tonnes and substitution of more than 244 lakh metric tonnes of crude oil underscore the strategic importance of bioresources in enhancing India's energy security. By substituting imported fossil fuels with domestically produced biofuels, India can reduce its trade deficit, buffer against volatile global oil prices, and lower its carbon footprint.

This dual benefit—boosting rural incomes while conserving foreign exchange—is a compelling argument for intensifying efforts on bioresources. But money apart, the program has contributed to avoiding about 700 lakh tonnes of CO_2 emissions, directly supporting India's climate commitments and its Net-Zero 2070 target. This reduction stems from ethanol's inherently lower carbon intensity compared to fossil fuels during combustion.

The ethanol industry is integrating cutting-edge technologies to improve production efficiency and sustainability. A key innovation is cellulosic technology, which produces ethanol from non-food biomass such as agricultural waste and energy crops through enzymatic hydrolysis and thermochemical processes. The sector is also exploring nanotechnology for enzyme immobilisation and designing more efficient catalysts. These advancements aim to make ethanol production more cost-effective while reducing greenhouse gas emissions. The use of leftover, damaged, or rotten food grains for ethanol production through fermentation processes is also actively pursued.

Apart from ethanol production, there are several other important processes for extracting bioenergy from biomass. Several different technologies, from simple combustion to generate heat energy, gasification, and fast pyrolysis, allow obtaining a wide range of useful products. Gasification converts biomass into a combustible gas called synthesis gas (syngas) through high-temperature partial oxidation, which can then be used for power generation or as a chemical feedstock. Pyrolysis heats biomass in the absence of oxygen, producing bio-oil, syngas, and charcoal, with bio-oil serving as a liquid fuel or chemical raw material. Anaerobic digestion breaks down organic materials without oxygen to produce biogas, mainly composed of methane, which can be used for heating, electricity, or transportation fuel. Additionally, transesterification transforms vegetable oils and fats into biodiesel, a renewable alternative to conventional diesel. These processes offer versatile ways to convert biomass into useful energy and fuels, each suited to different types of biomass and energy needs.

Integration with carbon capture and storage (CCS) technologies represents another crucial aspect of the modern syngas industry. By capturing the carbon emissions generated during gasification and fuel production, industries can significantly reduce their overall environmental impact. Research efforts are also focused on improving the efficiency, scalability, and cost-effectiveness of gasification processes, with innovative approaches such as plasma gasification and microbial conversion gaining attention. These advancements are expected to broaden the applications of syngas across various industries and accelerate the transition toward a low-carbon economy.

Looking ahead, the role of syngas in the modern era is poised to expand further as technological innovations continue to address current limitations. Its capacity to convert waste into valuable fuels and chemicals, coupled with ongoing developments in renewable energy systems and carbon reduction strategies, makes syngas a cornerstone of a sustainable energy future. As investments in research and infrastructure grow, syngas is likely to play an increasingly important role in reducing global dependence on fossil fuels, supporting environmental goals, and fostering a resilient, low-emission energy landscape worldwide.

At the heart of each of the technological processes are fundamentally different options for the chemical transformation of organic raw materials, determined primarily by the presence of oxygen or the nature of interaction with it. In the most uncompromising cases, the combustion of biomass occurs in the free presence of oxygen (air). A gamut of technologies is built around the availability of oxygen and temperature.

Torrefaction is slow, low-temperature pyrolysis, at around 200 °C, which effectively removes moisture and a significant part of volatiles from biomass, thereby increasing its calorific value and increasing embrittlement. In fast pyrolysis at 500 °C, in the absence of oxygen, and with a very short residence time of several seconds, about 75% of a bioenergy liquid called bio-oil can be obtained, stored or transported over a long distance.

Biomass to Gasoline via Biomass to Liquids (BTG-BTL) is a process that involves converting biomass into liquid fuels such as gasoline, diesel, and jet fuel. This technology combines gasification of biomass to produce syngas (a mixture of carbon monoxide and hydrogen), followed by further chemical refining, often through catalytic conversion, to produce high-quality liquid fuels compatible with existing infrastructure. BTG-BTL aims to provide renewable, sustainable alternatives to fossil fuels by utilising biomass waste or energy crops. It offers an efficient way to produce transportation fuels while reducing greenhouse gas emissions compared to conventional fossil fuel production.

Another transformative development is syngas in the field of renewable energy and chemical manufacturing. Syngas is a mixture of carbon monoxide, hydrogen, and other gases, produced primarily through gasification of biomass, coal, or waste materials. This process has revolutionised how we produce fuels and chemicals, enabling more sustainable and flexible production methods. It is not a new idea, but it has gained currency only recently.

German scientists Franz Fischer and Hans Tropsch developed in the 1920s a catalytic chemical process that converts syngas (a mixture of carbon monoxide and hydrogen) into liquid hydrocarbons, including fuels such as diesel, jet fuel, and other waxes. The process involves passing syngas over a catalyst, typically made of iron or cobalt, at elevated temperatures and pressures. Under these conditions, the carbon monoxide and hydrogen molecules chemically react to form long-chain hydrocarbons that can be refined into usable fuels.

In recent years, Fischer-Tropsch (FT) technology has gained renewed interest. Natural gas and coal are converted into cleaner liquid fuels, and various types of products can be obtained –

diesel fuel, paraffins, gaseous fuels, and wax. The integration of FT technology into the Escravos-Lagos pipeline, which transports natural gas from the Escravos region in the Niger Delta to Lagos, Nigeria's largest city and economic hub, which feeds power plants, industrial processes, and domestic consumption allows for the conversion of natural gas into cleaner liquid fuels. Another area of production of various fuels from syngas is the intermediate production of methanol and its derivatives, as well as methanation. A combination of temperature (220 °C to 300 °C) and pressure (50 to 100 bar) is used to obtain methanol from syngas.

Second-generation ethanol production refers to biochemical conversion methods based on enzymatic or microbiological processes. Unlike the alcoholic fermentation of biomass with a high sugar and starch content, for the conversion of cellulosic biomass, it is necessary to first isolate carbohydrates from cellulose and hemicellulose and only then ferment them into ethanol. Cellulose is a very stable substance that is difficult to destroy even by acid hydrolysis. Special enzymes have been developed, and the process itself is called enzymatic hydrolysis. Among the most popular fermentation methods are yeast fermentation, microbial fermentation via acetic acid, and microbial fermentation via farnesene, a naturally occurring organic compound classified as an olefin or terpene. In addition, methods of pre-treatment of cellulosic biomass have found wide application, including steam explosion, ozone oxidation, and ammonia treatment.

In 2015, DuPont commissioned one of the largest second-generation bioethanol plants in Iowa, USA, with a capacity of 30 million gallons per year. The technology includes pre-treatment of remnants of corn plant stems and leaves, grinding and additional splitting; then the first stage of fermentation is carried out by saccharification to break down complex carbohydrates into simple sugars for fermentation; then the next stage of fermentation by means of patented strains of microorganisms; finally, during distillation, ethanol vapours are captured and condensed into liquid ethanol.

Investing in research and development for advanced bio-refineries, creating a supportive policy environment, and establishing robust supply chains are essential steps. These efforts can make bioresources more economically viable and environmentally sustainable, enabling India to meet its climate commitments, reduce dependence on fossil fuels, and promote rural livelihoods. Without doubt, the emphasis on a broader range of bioresources can accelerate India's transition towards a sustainable, resilient, and diversified energy system, ensuring long-term economic growth and environmental health.

Unlike first-generation biofuels derived from food crops and second-generation biofuels from non-food biomass, such as lignocellulosic materials, third-generation biofuels focus on using algae due to their high oil content, fast growth rates, and ability to thrive in various environments. These microalgae can be cultivated in wastewater or non-arable land, making them a sustainable and efficient source of biofuels, including biodiesel, biojet fuel, and biogas. The development of third-generation biofuels aims to provide a large-scale, environmentally friendly alternative to fossil fuels while minimising competition with food production.

The advantages of this type of product are the possibility of obtaining significant volumes of biofuel in relatively small areas, even unsuitable for agricultural production; when growing algae, there is no acute need for fresh water; microalgae can contain a record high oil level – up to 60-80%; these plants are classified as fast-growing, and some species in favourable conditions can double their weight within a few hours; during their growth, microalgae assimilate carbon dioxide from the atmospheric air. Comparing the potential of oil yields, for example, the simplest microalgae gives twice or more the extraction rates of palm oil, which is around 650 gallons/acre.

There are three principal ways to cultivate microalgae: phototrophic cultivation, which involves light and CO_2; heterotrophic cultivation, where growth occurs in the absence of light but with the addition of a special organic substrate; and mixotrophic cultivation, where microalgae can grow under both phototrophic and heterotrophic conditions. The process of growing microalgae is carried out either in open (Open Raceway Pond, ORP) or closed systems like Photobioreactor (PBR). Closed tubular systems require a large illumination area, but there is good biomass productivity. Of course, it is limited by active fouling and the need to control gradients of pH, dissolved oxygen and CO_2 along the tubes.

No wonder that bioenergy from microalgae is considered the future of renewable energy due to its numerous advantages. Microalgae can produce high yields of oils, carbohydrates, and proteins, making them a versatile source for biofuels such as biodiesel, biojet fuel, and biogas. They require minimal land and freshwater compared to traditional crops, and they can grow rapidly in a variety of environments, including wastewater and saltwater, which helps avoid competition with food production. Additionally, microalgae capture CO_2 during their growth, contributing to carbon sequestration and reducing greenhouse gases. With ongoing advancements in cultivation, harvesting, and processing technologies, bioenergy from microalgae holds great potential to sustainably meet future energy demands while supporting environmental conservation efforts. Advances in genetic engineering, cultivation technology, and harvesting methods are expanding the commercial potential of microalgae biotechnology.

The coastal regions of India, particularly the Gulf of Kutch in Gujarat and Sundarbans in West Bengal, hold significant potential for microalgae cultivation for energy production. Sundarbans, the largest mangrove forest in the world, holds significant ecological and economic potential. The region's unique ecosystem, characterised by brackish water, high biodiversity, and abundant sunlight, provides an ideal environment for cultivating diverse microalgae species. These microorganisms can contribute to environmental sustainability by aiding in water purification, carbon sequestration, and nutrient cycling. Additionally, microalgae cultivation in the Sundarbans can generate valuable bio-products such as biofuels, nutraceuticals, and pharmaceuticals, providing new livelihood opportunities for local communities. Harnessing microalgae technology in this sensitive ecosystem also promotes eco-friendly development, helping to balance environmental conservation with economic growth in the region.

Similar to how oil unlocked economic growth in desert areas through resource extraction, microalgae hold the potential to bring prosperity to marshlands and wetlands. India's marshlands cover an estimated area of around 15.26 million hectares. By cultivating microalgae in these ecosystems, we can produce valuable biofuels, nutritional supplements, pharmaceuticals, and other bio-based products. This sustainable approach can stimulate local economies, create jobs, and promote environmental conservation, helping to revive and utilise marshlands productively. In essence, microalgae can serve as a green "oil" for marshlands, fostering ecological health and economic development simultaneously.

As this book draws to a close, one truth shines clear: villages are not remnants of the past, but the seedbeds of humanity's future. Gandhi dreamed of self-reliant Gram Swaraj, Kalam envisioned PURA empowered by science, and today NextGen Biology and Artificial Intelligence together make these dreams practical and urgent. The challenge is not merely to adopt technology, but to harness it ethically, inclusively, and sustainably so that it uplifts the weakest, empowers smallholders, and renews the dignity of rural life. In this lies the resurrection of the Village Republic—not as nostalgia, but as a living reality for the twenty-first century. Visions outlive visionaries; and when joined with the tools of their time, they rise again—not as dreams, but as destiny.

•••

EPILOGUE

Time proves that great visions never die. They return with new tools, new hands, and an eternal purpose— to uplift humanity.

Great visions are not bound by the life of the visionary. They outlive the mortal man, waiting patiently for their time. When the world is ready—when science has advanced, when technology matures, when humanity begins to seek answers anew—such visions rise again, no longer as dreams, but as living forces. Gandhi's Gram Swaraj and Kalam's PURA are such eternal visions.

Today, through the power of NextGen Biology and Artificial Intelligence, they find their moment. What was once an ideal has become achievable; what was once distant has become urgent. At the heart of this renewal lies the dawn of a bioresource-based economy—where crops, residues, bamboo, algae, microbes, and agro-waste are transformed into fuels, fibres, medicines, plastics, and energy. Waste becomes wealth, villages become value-creators, and sustainability becomes prosperity.

The Village Republic 2.0 is not a return to the past, but the rebirth of the future — powered by bioresources, guided by intelligence, and rooted in human dignity.

ACKNOWLEDGEMENTS

With utmost humility, I bow to **the Supreme Power**, the eternal source of inspiration, strength, and wisdom that has guided me through every step of my life and work.

My deepest gratitude is reserved for **Prof. Arun Tiwari**, my co-author, whose profound intellect, creative discipline, and unwavering friendship transformed this book from an idea into a living dialogue between science and society. His ability to find meaning at the intersection of biology, technology, and humanity has been both a gift and a guiding light.

I owe everything to my **wife, Richa**, whose patience, affection, and quiet strength have sustained me through every challenge, and to my beloved **children, Avisha and Vayun**, whose love and curiosity give me unending joy and purpose. My heartfelt gratitude also goes to my **parents**, who instilled in me the values of sincerity and service, and to my **extended family**—my brothers, sisters, and their families; and my wife's parents, brother, sisters, and their families—for their blessings, understanding, and constant support.

My special thanks go to the **project staff** who worked with me at various stages during my tenure at the **CSIR–Institute of Himalayan Bioresource Technology (CSIR-IHBT), Palampur, Himachal Pradesh**. Your dedication made every idea take root and bloom. I fondly acknowledge my former colleagues at CSIR-IHBT, including **Mr Pabitra Gain** for his artistry in photography, **Dr Rakesh K. Sud** for initiating linkages with rural Himachal, and **Dr Rakesh Kumar** for strengthening our connection with the North East. I extend special appreciation to **Dr Sukhjinder Singh** for fostering ties with entrepreneurs and other stakeholders, and to **Dr Mahesh Gupta** for strengthening linkages with start-ups and innovation-led ventures.

I extend my warm appreciation to my **colleagues from ICAR–Indian Agricultural Research Institute (ICAR-IARI)** and my **current team at the Agricultural Scientists Recruitment Board (ASRB), New Delhi**, for their cooperation and camaraderie. I also express sincere gratitude to all those who took time to contribute forewords and messages for this book—their words add richness and perspective to its pages.

My heartfelt thanks go to my **mentors, senior colleagues, friends, and well-wishers**, whose guidance and encouragement have shaped my journey. I also acknowledge my **foreign collaborators and friends**, who brought global insight and partnership to our shared pursuit of sustainable bioresource development.

This work owes much to the **publisher, Mr Ashutosh Ramgir**, and the **editorial team**, as well as to all who worked tirelessly behind the scenes to bring it to completion. I record my sincere appreciation for **Mr Ram Gopal Shukla,** who ensured smooth collaboration, diligent follow-through, and exemplary integrity throughout this project.

I express deep gratitude to the **farmers and communities** in **Himachal Pradesh, the North East, Uttarakhand, Punjab, and other parts of India**, whose trust and participation transformed science into livelihoods. My appreciation extends to **startups, entrepreneurs, and all stakeholders** who are helping to carry forward the spirit of innovation and self-reliance that this book advocates.

I gratefully acknowledge the **Council of Scientific and Industrial Research (CSIR)**, the **Indian Council of Agricultural Research (ICAR)**, the **Department of Agricultural Research and Education (DARE)**, and all **funding agencies** whose support made our research and outreach possible. To my **students and mentees**, I remain indebted—you are the true continuation of this journey. To my **teachers and professors**, who shaped my scientific temperament and values, I owe a lifelong debt of gratitude.

And finally, to **the readers of this book**—thank you for lending your time and mind to these ideas. May *Village Republic 2.0* ignite thought, inspire action, and reaffirm our shared faith in science as a force for human progress and planetary harmony.

— **Dr Sanjay Kumar**
Palampur / New Delhi

ADVANCE PRAISE FOR THE BOOK

Village Republic 2.0 conveys a simple yet powerful message: farmer prosperity will come from value addition and the wise application of modern tools, such as NextGen Biology and AI. The path to a self-reliant India runs through its villages, guided by innovation, enterprise, and the enduring spirit of its people.

Dr Mangi Lal Jat,
Secretary to the Government of India (DARE) and Director-General, ICAR

The central argument of *Village Republic 2.0* is both simple and powerful: India's villages are abundant in bioresources, and modern science, particularly NextGen Biology and artificial intelligence, can transform these into sources of wealth and well-being. Crops, residues, bamboo, algae, microbes, and agro-waste—often considered by-products—are reimagined here as raw materials for biofuels, bioplastics, nutraceuticals, and high-value biomaterials. This perspective shifts the narrative from subsistence to enterprise, from survival to prosperity. Artificial intelligence adds another dimension to this transformation.

Dr N. Kalaiselvi,
Director-General,
CSIR and Secretary, DSIR, Government of India

The book *Village Republic 2.0: Harnessing Bioresources through NextGen Biology and Artificial Intelligence* charts a compelling and forward-looking journey of rural transformation. It demonstrates how scientific innovation and technological integration, when rooted in equity and equality, can make a lasting impact. These forces must—and will—play a central role in shaping Our Common Future.

Dr Himanshu Pathak,
Director-General,
International Crops Research Institute for the
Semi-Arid Tropics (ICRISAT), Hyderabad

The book articulates a forward-looking framework where cutting-edge science, value addition and ecological prudence converge to uplift rural communities. Its strength lies in showing how modern biology and artificial intelligence can become instruments of empowerment, enabling farmers to rise as entrepreneurs and innovators. It reinforces the conviction that our villages are not just the custodians of tradition but also the crucibles of future growth.

Dr Trilochan Mohapatra,
Former Secretary to the Government of India (DARE) and
Former Director-General of ICAR

Bringing out the philosophy of Mahatma Gandhi and combining it with the Vision of Dr Abdul Kalam to exhibit the strength of Biosciences with new AI technologies for transforming the rural economy, the book gives a detailed account of the potential of new NextGen Biology technologies and the value-added products which can be produced to transform the rural economy. The message of the book is clear: new technologies in biosciences, combined with the latest tools of AI, have the potential to drive the economy and transform villages today.

Dr Renu Swarup,
Former Secretary to the Government of India,
Department of Biotechnology

I am convinced that, indeed, if a concerted effort is made to promote rural livelihoods through modern science, technology, and innovation, the rate of urban migration can be reduced, and it can prove beneficial to the country's overall development. The second part of the book outlines a clear vision of the future, which is articulated well.

Dr Shekhar C. Mande,
Biologist,
Former Secretary to the Government of India (DSIR) and
Former Director-General of CSIR

Not just a theoretical treatise, but a call to action — advocating for public sector-led innovation that empowers smallholders to transform from passive commodity producers into active stakeholders and value-creators in the bioeconomy. This book offers a future-ready blueprint for transforming villages into laboratories of innovation, inclusion, and resilience.

Prof. Sudhir K. Sopory,
Emeritus Senior Scientist,
ICGEB, and Former Vice Chancellor, Jawaharlal Nehru University

India's villages, being custodians of all resources, hold significant potential to drive the nation's inclusive growth through modernised agriculture, robust rural industries, digital connectivity, sustainable energy, and empowered local governance. Such developed villages can be christened as Smart Villages. Since most Indians reside in villages, developing Smart Villages should precede the Development of Smart Cities. India needs more Smart Villages than Smart Cities.

Prof. S.K. Sharma,
Geneticist, Former Vice Chancellor,
HP Agricultural University, Palampur

The pursuit of excellence at the individual level is intricately linked to achieving excellence at the institutional and broader societal levels. Linking science with society is a pressing need of the hour, given the increasing aspirations and widening income inequalities. "The villages are not a relic of the past but a laboratory of the future." May this feeling grow, spread, and fertilise the imagination and engagement of future scientists with society, no less.

Dr Anil Gupta,
Founder of Honey Bee Network, and
Former Professor at IIM, Ahmedabad

Born in the Himalayas and having worked on ecology, mountain resilience, and sustainable development for over 45 years, I have always been struck not by what the region lacked, but by what it quietly held—extraordinary biodiversity, traditional knowledge, rich ethnic diversity, and a resilient spirit. This book is ultimately a journey of hope and imagination. It will engage thinkers, doers, and young leaders in shaping a better future—one where our villages become hubs of innovation, resilience, and self-reliance.

Dr Eklabya Sharma,
Ecologist, and Former Director,
International Centre for Integrated Mountain Development (ICIMOD)

This book rightly places farmers, women, and youth at the centre of this transformation. It reminds us that technology is only meaningful when it uplifts people and strengthens community life. The "Village Republic" can indeed become the foundation of a prosperous, self-reliant, and sustainable India.

Dr Anil Prakash Joshi,
Green activist and Founder,
Himalayan Environmental Studies and
Conservation Organisation (HESCO)

The book provides both visionary and practical insights into rural development. Leveraging bioresources through NextGen Biology and artificial intelligence offers a compelling vision for how these tools can be used not only to modernise agriculture but also to rebuild rural economies on stronger, more resilient foundations.

Dr P.G. Rao,
Former Director, CSIR-NEIST, and Former Vice Chancellor,
University of Science and Technology, Meghalaya

Our villages have always been the heart of our nation. No longer limited to crop cultivation alone, agriculture is now emerging as a dynamic engine for prosperity, social well-being, and environmental regeneration. By harnessing these powerful technologies, villages can evolve into vibrant ecosystems of innovation where knowledge, enterprise, and ecological stewardship are integrated. *Village Republic 2.0* points us toward a future where rural India once again becomes the foundation of national prosperity and global sustainability.

Dr Rajendra Dobhal,
Vice Chancellor,
Swami Rama Himalayan University

The emergence of newer product development and bioprocessing technologies has opened pathways for addressing critical challenges such as food security and malnutrition. By transforming bio-resources into value-added products, India can not only empower its rural communities but also contribute significantly to building a healthier and more self-reliant nation. By weaving together the ideals of self-reliance, dignity, inclusivity, and technological innovation, the book provides a transformative vision for reimagining India's villages—not as remnants of the past, but as dynamic centres of prosperity, sustainability, and human progress.

Prof. Gauri Dutt Sharma
Vice Chancellor,
University of Science and Technology, Meghalaya

Village Republic 2.0 is a timely catalyst to transform the thinking and lives of people in India to consider villages as the target of future sociological experiments for boosting both the economy and the quality of life. By paving the way for reverse migration from cities to villages, it spearheads the movement to empower villages by promoting knowledge, as well as the circular economy, through the adoption of green and sustainable technologies. I sincerely believe that this book will accelerate the realisation of Gandhi's and Kalam's dreams sooner rather than later.

Dr Anil Tripathi,
Director,
Indian Institute of Science Education and Research Mohali

Village Republic 2.0 is much more than a book—it is a fervent call to reimagine India's rural future. By integrating Bapu's dream of Gram Swaraj with Dr Kalam's vision of PURA (Providing Urban Amenities in Rural Areas), the work shows that villages need not be symbols of deprivation, but can instead be centres of innovation, prosperity, and dignity. Its pages blend Sanjay's personal journey in science with Arun Tiwari's legacy of carrying forward Dr Kalam's dream, offering both intellectual depth and practical hope.

Dr Madhukar Garg,
President,
Refining and Petrochemicals R&D of Reliance Industries Limited
and Former Director-General, CSIR

I am confident that this book will inspire fresh thinking among policymakers, researchers, entrepreneurs, and, above all, the youth of India, to build a rural economy that is prosperous, resilient, and future-ready. It is a clarion call for harnessing science and technology in the service of human dignity and national transformation.

Dr H.P. Singh
Chairman,
Confederation of Horticulture Associations of India and
Former Deputy Director-General (Horticulture), ICAR

Scientific breakthroughs attain their fullest significance only when they permeate grassroots communities. Our villages are repositories of rich bioresources. The concept of a bioresource-based economy, driven by NextGen Biology and artificial intelligence, holds the potential not only to generate economic prosperity but also to diminish migration, rejuvenate rural ecosystems, and foster a renewed sense of pride within India's villages.

Dr V.M. Tiwari,
Director,
CSIR-North East Institute of Science and Technology, Jorhat.

For those working closely with farmers, it is evident that the future of Indian agriculture lies in engaging youth in it, and agriculture should be a lucrative and prestigious enterprise; ironically, this has eroded over time. Modern science and technology offer numerous opportunities to attract youth to villages. This book highlights exactly that—how modern biology and Artificial Intelligence can empower smallholders, create rural enterprises, and restore dignity to village life.

Dr Ashok Kumar Singh,
Vice Chancellor,
Rani Lakshmi Bai Central Agricultural University, Jhansi

Village Republic 2.0 is a work of deep contemporary relevance, addressing the aspirations of scientists, educators, farmers, policymakers, and young minds eager to shape India's future. It powerfully underscores that the path to a self-reliant India begins with healthy soils, strong biodiversity, empowered farmers, and vibrant villages.

Dr J.C. Katyal,
Former Vice Chancellor,
Chaudhary Charan Singh Haryana Agricultural University,
and Former Deputy Director General (Education), ICAR

There is a need to bring in transformative change(s) in the rural economy through the application of newer technologies, which are capable of not only making the villages self-reliant but also imparting much-needed dignity among the rural population in general and farmers in particular. The introduction of the circular economy, in conjunction with NextGen Biology and AI, will help trigger inclusive growth and overall prosperity in the country's villages during AMRIT KAAL.

Dr H.S. Gupta,
Former Director-General,
Borlaug Institute for South Asia,
and Former Director, Indian Agricultural Research Institute

Interweaving the intricate physiological processes with his findings of how plants survive under the harsh weather of cold deserts makes fascinating reading. The book lays the path for a bio-resource-based economy—where plants like '*Ratanjot*', apples, bamboo, algae, fungi, and microbes, and their products like flowers, perfumes, fibres, bioplastics, bioenergy, and medicinal molecules can transform villages into prosperous republics of the future.

Prof. Anupam Varma,
Virologist and Former ICAR - National Professor

Realising the dreams of the "Father of the Nation" and the "People's President" will require actionable plans for elevating India's villages onto a path of sustainable prosperity through innovative technologies. Given the scientific acumen of both the authors and their concern for society and the nation as a whole, I have a strong conviction that *Village Republic 2.0* will prove a turning point in harnessing science and technology for an *Atmanirbhar* and *Vikasit Bharat* through the development of its villages.

Dr M.S. Chauhan,
Vice Chancellor,
G.B. Pant University of Agriculture and Technology, Pantnagar

Books by Arun Tiwari

Published by Sakal Publications

- The Baton Should Not Fall
- Decoding the Pandemic
- India Wakes : Post Coronavirus New World Order
- Innovate Locally to Win Globally
- Live for a Legacy
- Uddhava Gita
- Discovery of Anugita
- Simple Spirituality : Recalling Kabir
- A Modern Interpretation of Tulsidas Ramcharitramanas
- A Modern Interpretation of Lokmanya Tilak's Gita Rahasya
- We are What We Eat
- Yoga Vasishtha
- Spectrum
- Purpose of Life
- Neo-ethical Living